Liquid–Solid Separators

There are no such things as applied sciences,
only applications of science.
Louis Pasteur (11 September 1871)

Dedicated to my wife, Anne, without whose unwavering support, none of this
would have been possible.

Industrial Equipment for Chemical Engineering Set

coordinated by
Jean-Paul Duroudier

Liquid–Solid Separators

Jean-Paul Duroudier

ELSEVIER

First published 2016 in Great Britain and the United States by ISTE Press Ltd and Elsevier Ltd

ISTE Press Ltd
27-37 St George's Road
London SW19 4EU
UK

www.iste.co.uk

Elsevier Ltd
The Boulevard, Langford Lane
Kidlington, Oxford, OX5 1GB
UK

www.elsevier.com

Notices

Knowledge and best practice in this field are constantly changing. As new research and experience broaden our understanding, changes in research methods, professional practices, or medical treatment may become necessary.

Practitioners and researchers must always rely on their own experience and knowledge in evaluating and using any information, methods, compounds, or experiments described herein. In using such information or methods they should be mindful of their own safety and the safety of others, including parties for whom they have a professional responsibility.

To the fullest extent of the law, neither the Publisher nor the authors, contributors, or editors, assume any liability for any injury and/or damage to persons or property as a matter of products liability, negligence or otherwise, or from any use or operation of any methods, products, instructions, or ideas contained in the material herein.

For information on all our publications visit our website at http://store.elsevier.com/

British Library Cataloguing-in-Publication Data
A CIP record for this book is available from the British Library
Library of Congress Cataloging in Publication Data
A catalog record for this book is available from the Library of Congress
ISBN 978-1-78548-182-6

Printed and bound in the UK and US

Contents

Chapter 3. Theory of Dialysis-Biological
Membranes – Electrodialysis   101

Preface

The observation is often made that, in creating a chemical installation, the time spent on the recipient where the reaction takes place (the reactor) accounts for no more than 5% of the total time spent on the project. This series of books deals with the remaining 95% (with the exception of oil-fired furnaces).

It is conceivable that humans will never understand all the truths of the world. What is certain, though, is that we can and indeed must understand what we and other humans have done and created, and, in particular, the tools we have designed.

Even two thousand years ago, the saying existed: "faber fit fabricando", which, loosely translated, means: *"c'est en forgeant que l'on devient forgeron"* (a popular French adage: *one becomes a smith by smithing*), or, still more freely translated into English, "practice makes perfect". The "artisan" (faber) of the 21st Century is really the engineer who devises or describes models of thought. It is precisely that which this series of books investigates, the author having long combined industrial practice and reflection about world research.

Scientific and technical research in the 20th century was characterized by a veritable explosion of results. Undeniably, some of the techniques discussed herein date back a very long way (for instance, the mixture of water and ethanol has been being distilled for over a millennium). Today, though, computers are needed to simulate the operation of the atmospheric distillation column of an oil refinery. The laws used may be simple statistical correlations but, sometimes, simple reasoning is enough to account for a phenomenon.

Since our very beginnings on this planet, humans have had to deal with the four primordial "elements" as they were known in the ancient world: earth, water, air and fire (and a fifth: aether). Today, we speak of gases, liquids, minerals and vegetables, and finally energy.

The unit operation expressing the behavior of matter are described in thirteen volumes.

It would be pointless, as popular wisdom has it, to try to "reinvent the wheel" – i.e. go through prior results. Indeed, we well know that all human reflection is based on memory, and it has been said for centuries that every generation is standing on the shoulders of the previous one.

Therefore, exploiting numerous references taken from all over the world, this series of books describes the operation, the advantages, the drawbacks and, especially, the choices needing to be made for the various pieces of equipment used in tens of elementary operations in industry. It presents simple calculations but also sophisticated logics which will help businesses avoid lengthy and costly testing and trial-and-error.

Herein, readers will find the methods needed for the understanding the machinery, even if, sometimes, we must not shy away from complicated calculations. Fortunately, engineers are trained in computer science, and highly-accurate machines are available on the market, which enables the operator or designer to, themselves, build the programs they need. Indeed, we have to be careful in using commercial programs with obscure internal logic which are not necessarily well suited to the problem at hand.

The copies of all the publications used in this book were provided by the *Institut National d'Information Scientifique et Technique* at Vandœuvre-lès-Nancy.

The books published in France can be consulted at the *Bibliothèque Nationale de France*; those from elsewhere are available at the British Library in London.

In the in-chapter bibliographies, the name of the author is specified so as to give each researcher his/her due. By consulting these works, readers may gain more in-depth knowledge about each subject if he/she so desires. In a reflection of today's multilingual world, the references to which this series points are in German, French and English.

The problems of optimization of costs have not been touched upon. However, when armed with a good knowledge of the devices' operating parameters, there is no problem with using the method of steepest descent so as to minimize the sum of the investment and operating expenditure.

Filtration of Liquids and
Gases Screening of Divided Solids

1.1. General considerations

1.1.1. *Fundamental parameters [DUR 99]*

From a filtration point of view, a divided solid is characterized by its specific resistance or its permeability.

Specific resistance is expressed by the equation:

$$\alpha = \frac{4\sigma^2(1-\varepsilon)}{\rho_s \varepsilon^3} \qquad \left(\text{m.kg}^{-1}\right)$$

ε is the porosity of the cake that accumulates on the filter medium, i.e. the volumetric void fraction of the filter cake.

The porosity of the dust suspended in a gas is of the order of 0.7. That of a bed of crystals is about 0.4 if the particles are more or less equidimensional. The porosity of platelets is around 0.8 and that of a fiber mat is greater than 0.9. Porosity increases artificially if the particles have concave areas on their surfaces. It is better not to take account of these hollows when calculating ε.

σ is the mean volumetric surface of the solid particles.

– sphere of diameter d $\sigma = 6/d$

– plate of thickness e $\sigma = 2/e$

– long cylinder with a circular cross-section of diameter d $\sigma = 4/d$

If the particles are different in shape and type, the mean value of σ is:

$$\sigma = \frac{\sum_i \frac{m_i \sigma_i}{\rho_{si}}}{\sum_i \frac{m_i}{\rho_{si}}}, \text{ where } \sum m_i = 1$$

m_i: gravimetric fraction of the particle i

ρ_{si}: actual density of the particle i: $kg.m^{-3}$

More generally, solid particles are classified into three main sizes:

– maximum;

– intermediate;

– minimum.

Classification by screening is based on the intermediate size, whereas the volumetric surface should be based on the minimum size. In common divided solids, the ratio of these two sizes can vary from 2 to 5, even if we discount platelets. In the latter case, the ratio can be as much as 1000 or more.

It would therefore be wise to write:

$$\sigma = \frac{6}{d_{min}}, \text{ where } \frac{1}{d_{min}} = \frac{\sum_j \frac{m_j}{d_{min.j}}}{\sum_j m_j}$$

Thus, d_{min} would be the harmonic mean of the $d_{min.j}$ with $_i$

$$\frac{d_{intermediate}}{d_{min}} = K \text{ independent of } j$$

Indeed, in practice, σ is often under-evaluated, which is why we chose d_{min}.

Specific Cake Formation (SCF) is the ratio of the solid mass M_S of the filter cake to the volume Ω_{Fi} of the filtrate having passed through the filter cloth. This volume is obtained after removal of the saturation liquid from the cake. This notion holds true as long as the cake remains saturated with the pre-filter solution.

We shall calculate SCF in a manner slightly different to that used in Duroudier [DUR 99].

The pre-filter solution contains c kilograms of solid per m^3 of *pure* liquid. When the mass M_s of solid is retained by the filter cloth, liquid is also retained, as it occupies the voids within the cake. The porosity of this cake is ε.

The volume of the filter cake is:

$$\frac{M_s}{\rho_s (1-\varepsilon)}$$

M_s: mass of the cake

ρ_s: density of the solid

The filter cake M_s is formed by:

– the pre-filter solution Ω at the concentration c

– the dispersion previously occupying the volume of the cake

$$M_s = c\Omega + c\frac{M_s}{\rho_s (1-\varepsilon)}$$

The Specific Cake Formation is:

$$SCF = G = \frac{M_s}{\Omega_{Fi}} = \frac{c}{1 - \dfrac{c}{\rho_s(1-\varepsilon)}}$$

In the expression of G, we can see that, for a gas:

$$G = c\,(\text{since } c \ll \rho_s)$$

The resistance of the filter medium – generally a cloth – is highly dependent on the nature of the fluid: a stronger cloth is required to filter a liquid than to filter a gas.

For a liquid $0.5 \times 10^9\ m^{-1} < R < 5 \times 10^9\ m^{-1}$

For a gas $10^7\ m^{-1} < R < 10^8\ m^{-1}$

Grace [GRA 56] reviewed the properties of filter media and environments. Kaplan and Hsu [KAP 79] studied the pressure drop across a fiber mat.

The viscosity of the liquid or gas is a determining factor in calculating the pressure drop of the fluid.

Ambient air $\mu \# 20 \times 10^{-6}$ Pa.s

Water at 25°C $\mu \# 10^{-3}$ Pa.s

It is often accepted that dissolving a solute in water doubles the viscosity of the latter, but there are many exceptions to this "law". Solutions of sugar in water are very viscous.

Remember that:

$$1\ cp = 1 \text{ centipoise} = 10^{-3}\ \text{Pa.s}$$

1.1.2. *Pressure drop across a bed of spherical particles (Ergun, 1952)*

This problem was studied by Carman [CAR 37], and examined anew by Ergun [ERG 52].

Such a bed is porous. Let us liken each pore to a hollow cylinder, at least to start with. All of the pores are considered to be identical.

The law governing the pressure drop ΔP in a tube is written as follows [BRU 68]:

$$\frac{\Delta P}{L_o} = \frac{\psi}{D_o} \times \frac{1}{2} \rho V_o^2, \quad \text{where} \quad \psi = \frac{64}{Re} + 0.77 = \frac{64\mu}{\rho V_o D_o} + 0.77$$

L_o and D_o are the length and the diameter of the tube: m

V_o is the flowrate of the fluid in the tube: $m.s^{-1}$

ρ and μ are the density and the viscosity of the fluid: $kg.m^{-3}$ and Pa.s

Therefore:

$$\frac{\Delta P}{L_o} = \frac{\rho V_o^2}{2} \left(\frac{64\mu}{\rho V_o D_o^2} + \frac{0.77}{D_o} \right)$$

However, (see [DUR 99]):

$$\frac{1}{D_o} = \frac{3(1-\varepsilon)}{2 \times \varepsilon d_p} \quad \text{and} \quad \frac{1}{D_o^2} = \frac{9(1-\varepsilon)^2}{4\varepsilon^2 d_p^2}$$

ε is the porosity (void fraction) of the bed which is passed through

d_p is the diameter of the particles (considered to be spherical): m

Therefore:

$$\frac{\Delta P}{L_o} = \frac{32\mu V_o 9(1-\varepsilon)^2}{4d_p^2 \varepsilon^2} + \frac{0.77 \times 3 \; \rho V_o^2 (1-\varepsilon)}{4d_p \times \varepsilon}$$

Let us posit that [DUR 99]:

$$L_o = tZ \qquad\qquad V_o = \frac{Vt}{\varepsilon}$$

Z: thickness of the particle bed (m)

V: velocity of the fluid in an empty container (m.s^{-1})

t: tortuosity of the particle bed (t = 1.45)

Thus:

$$\frac{\Delta P}{Z} = \frac{32 \times 9 \ V(1-\varepsilon)^2 \times 1.45^2}{4 \ d_p^2 \varepsilon^2} + \frac{0.77 \times 3 \ \rho V^2 (1-\varepsilon) \times 1.45^3}{4 d_p \varepsilon^3}$$

$$\frac{\Delta P}{Z} = \frac{151.4 \ V(1-\varepsilon)^2}{d_p^2 \varepsilon^3} + \frac{1.76 \ \rho V^2 (1-\varepsilon)}{d_p \varepsilon^3}$$

Ergun [ERG 52] established this equation by simplifying the coefficients slightly, and by replacing 151.4 with 150, and 1.76 with 1.75.

NOTE.–

When the particles are similar in shape, but not spherical, we need to multiply their diameter d$_p$ by the sphericity factor ϕ_s.

$$\phi_s = \frac{\text{surface area of a sphere of the same volume}}{\text{surface area of the particle}}$$

NOTE.–

B.H. Kaye [KAY 67] described a method for deducting the value of the volumetric surface area of the particles from permeability value.

NOTE.–

S.J. Kaplan and C.D. Morland [KAP 79] proposed a method for determining the pressure drop of a non-Newtonian liquid across a bulk fiber mat.

Sater and Levenspiel [SAT 66] studied the gas–liquid flow in a divided-solid bed.

NOTE.–

The specific resistance is [DUR 99]:

$$\alpha = \frac{4\sigma^2 (1-\varepsilon)}{\rho_s \varepsilon^3} \qquad\qquad [1.1]$$

σ is the volumetric surface of the solid bodies:

sphere: (of diameter d) $\qquad \sigma = \dfrac{\pi d^2}{\pi d^3 / 6} = \dfrac{6}{d}$; plate: (of thickness e) $\qquad \sigma = \dfrac{21}{le} = \dfrac{2}{e}$

NOTE.–

The permeability – so crucially important to geologists and those working on foundations – is:

$$K = \frac{1}{\alpha \rho_s (1-\varepsilon)} = \frac{\varepsilon^3}{4\sigma^2 (1-\varepsilon)^2} = \frac{\varepsilon^3 d^2}{150 (1-\varepsilon)^2} \qquad \left(m^2\right)$$

[KAY 67] used the permeability value to derive the total surface area of the particles.

1.1.3. *General laws of filtration*

The fundamental law, known as Darcy's Law, is stated as follows:

$$\Delta P = \mu \frac{d}{d\tau}\left(\frac{\Omega}{A}\right)\left[R + \frac{M_s}{A}\alpha\right] = \mu \frac{d}{d\tau}\left(\frac{\Omega}{A}\right)\left[R + \alpha G \frac{\Omega}{A}\right]$$

1) Integration at constant flowrate ($\Omega = Q\tau$):

$$\Delta P = \mu \frac{Q}{A}\left[R + \alpha G \tau \frac{Q}{A}\right]$$

The pressure drop increases linearly over time.

2) Integration at constant pressure drop:

$$\tau = \frac{\mu\Omega}{\Delta PA}\left[R + \frac{1}{2}\alpha G\frac{\Omega}{A}\right]$$

or else:

$$\frac{\Omega}{A} = \frac{-R + \sqrt{R^2 + \frac{2\alpha G\Delta P\tau}{\mu}}}{\alpha G} \tag{1.2}$$

Note that these laws only hold as long as the cake is completely immersed in the liquid. The next step is to evacuate the liquid, i.e. to purge the cake.

1.1.4. *Purging the cake (Duroudier, 1999)*

We shall distinguish two main cases:

1) Purging on a filter used at constant ΔP:

The purge time, for an initially saturated filter cake, is [DUR 99]:

$$\tau_{PG} = \frac{\mu_L \varepsilon(1 - S_I)}{\Delta P - (1 - \varepsilon)\sigma\gamma}\left(\frac{Z_o^2}{2K} + RZ_o\right)$$

Z_o: initial cake thickness (m)

S_I: irreducible saturation

$$S_I = \frac{\text{volume of liquid}}{\text{volume of interparticle voids}}$$

γ: surface tension of the liquid (N.m^{-1})

A case study [DUR 99] showed that this purge time seldom exceeds one-tenth of a second. However, this was never compared against actual experience.

The irreducible saturation is due to the menisci that form between adjacent particles, or to the liquid in the concavities on the surface of the particles. This saturation represents, at most, 10% of the porosity.

Wakeman [WAK 82] determined the variation in saturation and gas flowrate as a function of time using a series of curves. Each curve is characterized by the difference in pressure across the cake. The relative permeabilities used by the author are justified in section 4.8.2 in Volume 4. The author found a good correlation between his results and actual experience.

2) Purging in a batch centrifuge:

The filter-cake purge time is [DUR 99]:

$$\tau_{PG} = \frac{R\mu_L\varepsilon(1-S_I)}{\rho_L r_p^2 \omega^2} Ln\left[\frac{u_1 + E}{u_2 + E}\right]$$

where:

$$E = \frac{2(1-\varepsilon)\sigma\gamma}{\rho_L\omega^2 r_p^2} \quad \text{and} \quad u_1 = \frac{r_p^2 - r_o^2}{r_p^2}; \quad u_2 = \frac{r_p^2 - r_2^2}{r_p^2}$$

r_p: radius of the drum

r_o: radius of the inner face of the cake

r_2 is given by:

$$(r_p - r_2) = 0.01(r_p - r_o)$$

Whilst the cake-formation time is less than one minute, the drainage (purge) time is ten times greater. It is therefore the purge time that determines the duration of the filtering operation.

3) Irreducible moisture of the filter cake:

This is the ratio of liquid mass to dry-solid mass:

$$X_I = \frac{S_I\varepsilon\rho_L}{(1-\varepsilon)\rho_s}$$

1.1.5. *Equivalent pore: a useful concept*

When a given number of spherical particles are packed together, we obtain a porous medium whose pore diameter can be calculated [DUR 99].

$$d_{pore} = \frac{4\varepsilon}{\sigma(1-\varepsilon)}; \text{ where } \sigma = \frac{6}{d}$$

This equation does not hold for fibers, and probably not for platelets and flakes.

EXAMPLE.–

By sintering a set of 1 μm spherical particles, we obtain a porous body with a porosity of 0.35.

$$d_{pore} = \frac{4 \times 0.35 \times 10^{-6}}{(1-0.35)6} = 0.35 \times 10^{-6} \text{ m} = 0.35 \text{ μm}$$

These 1 μm spherical particles block the passage of all particles whose diameter is one-third of that of the spherical particles.

1.2. Liquid filters

1.2.1. *Partially immersed filter cloth*

A drum filter rotating about a horizontal axis is the most common type of device.

The cloth is immersed in the pre-filter solution so that the latter covers 20% of its surface area. The purged cake is then washed in a spray of wash liquor, which is then vacuumed off. Compressed air is blown through the cloth, causing the cake to become partially detached before being scraped off by a blade.

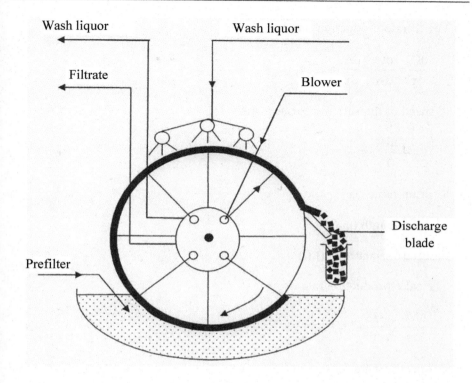

Figure 1.1 *Drum filter*

The filter operates at constant pressure drop. The law of filtration is stated as follows:

$$\frac{\Omega}{A} = \frac{-R + \sqrt{R^2 + \dfrac{2\alpha G \Delta P \tau_I}{\mu}}}{\alpha G} = f(\tau_I)$$

τ_I: immersion time (s)

T: drum rotation period (s)

η_I: immersion ratio

$$\tau_I = T\eta_I$$

The filtrate-production rate is:

$$\frac{d\Omega}{d\tau} = \frac{d\Omega}{dA} \times \frac{dA}{d\tau} = vf(\tau_I)$$

v: speed of the filtering surface $(m^2.s^{-1})$

$$v = \frac{2\pi RL}{T} = \frac{S}{T}$$

R: drum radius (m)

L: drum length (m)

S: drum surface area (m^2)

The cake-production rate is:

$$\frac{dM_s}{d\tau} = \frac{GS}{T}f(\tau_I) \quad (kg.s^{-1})$$

G: specific cake formation $(kg.m^{-3})$

EXAMPLE.–

We want to calculate the production rate of a paraffin filter in a refinery. Paraffins crystallize as platelets with a thickness of 10^{-7} m.

c = 135 kg.m^{-3}	η_I = 0.20	ΔP = 0.3.10^5 Pa
ρ_s = 900 kg.m^{-3}	T = 60 s	μ = 5.10^{-3} Pa.s
A = 40 m^2	ε = 0.8	R = 0.4×10^9 m^{-1}

$$\sigma = \frac{2}{10^{-7}} = 2 \times 10^7 \, m^{-1}$$

$$\alpha = \frac{4 \times 4.10^{14}(1-0.8)}{900 \times 0.8^3} = 6.94 \times 10^{11} \, m.kg^{-1} \quad (\text{equation } [31.1])$$

$$G = \frac{135}{1 - \dfrac{0.8 \times 135}{900 \times 0.2}} = 337.5 \text{ kg.m}^{-3}$$

$$\tau_I = 0.2 \times 60 = 12 \text{ s}$$

According to equation [1.2]:

$$\frac{\Omega}{A} = \frac{-0.4 \times 10^9 + \sqrt{0.16 \times 10^{18} + \dfrac{2 \times 337.5 \times 6.94 \times 10^{11} \times 12 \times 0.3 \times 10^5}{5 \times 10^{-3}}}}{337.5 \times 6.94 \times 10^{11}}$$

$$\frac{\Omega}{A} = 0.000784 \text{ m}^3.\text{m}^{-2}$$

$$\frac{dM_s}{d\tau} = \frac{337.5 \times 40 \times 0.000784}{60} = 0.176 \text{ kg.s}^{-1}$$

NOTE.–

In this example, the thickness of the paraffin platelets was not calculated directly, but rather was derived from the specific resistance using a temperature-controlled Büchner filter (see [DUR 99]).

1.2.2. Projecting the pre-filter solution on to a belt filter cloth

The pre-filter solution is allowed to flow on to the belt filter through a vent that is perpendicular to the axis of progression of the belt. The flowrate of the pre-filter solution per meter of belt width is q_F ($m^2.s^{-1}$). The rate of progression of the belt is v_B ($m.s^{-1}$).

The initial height H of the pre-filter solution is:

$$H = Q/v_B$$

The problem corresponds to the analysis of a non-continuous filtration process, and, to determine the required belt length, we only need to determine the filtration time and the cake purge time, given that the device is operating at constant ΔP.

On the surface of the belt element dA:

– the solid mass is M_s

– the cake height to be obtained is:

$$H_G = \frac{M_s}{\rho_s (1-\varepsilon)}$$

– the free-liquid height to be transformed into the filtrate is:

$$H_L = H - H_G = H\left(1 - \frac{c}{(1-\varepsilon)\rho_s}\right)$$

In the equation for the filtration at constant ΔP, we have:

$$\frac{\Omega}{A} = H_L = \frac{Q}{v_B}\left(1 - \frac{c}{(1-\varepsilon)\rho_s}\right)$$

Lastly, the filtration time, without purging of the cake, is:

$$\tau_{fil} = \frac{\mu H_L}{\Delta P}\left[R + \frac{1}{2}\alpha G H_L\right]$$

Moreover, we know [DUR 99] that the cake purge time is:

$$\tau_{PG} = \frac{\mu\varepsilon(1-S_1)}{\Delta P - (1-\varepsilon)\sigma\gamma}\left(\frac{H_G^2}{2K} + R H_G\right)$$

K: permeability of the cake (m^2)

The belt length is obviously:

$$L_B = v_B \left(\tau_{fil} + \tau_{PG} \right)$$

1.2.3. *Choice of liquid filter*

In the chemical- and food industries, the most commonly used filter, by far, is the vacuum drum filter.

However, if the cake is not able to remain attached to the filter cloth when the latter is vertical, a horizontal cloth must be used.

In this case, we can choose one of the following:

– a horizontally rolling belt filter;

– a bucket filter (emptied by tipping);

– a horizontal, circular-plate filter.

However, standard filtration is impossible if:

$$\alpha\mu \geq 2.5 \times 10^9 \, s^{-1}$$

In this case, we would need to consider:

– filtration at a pressure of a few bars, provided the cake is not compressible;

– centrifugal decantation, if the cake is compressible.

Deep-bed filtration is particularly well-suited for very-dilute solutions (less than 2% of solid per volume).

For the removal of trace solids, a cartridge filter may be necessary.

Ultra-fine suspensions ($d_p < 0.5 \, \mu m$) may be flocculated beforehand, but if the flocks are too large, clogging may occur. Press-filter filtration is preferred with low levels of flocculation as it minimizes the passage of fine particles through the filter cloth.

1.3. Centrifugal spin-drying (liquid–solid suspensions)

1.3.1. *Theoretical background*

Our description of this process will be different to (and clearer than) the description proposed by this author in 1999. The new approach comprises four steps:

1) Check the initial fill rate of the basket:

The volume of the basket is:

$$\Omega_p = \pi r_p^2 H$$

r_p: basket radius (m)

H: basket height (m)

The suspension volume is:

$$\Omega_{sus} = \frac{M_s}{c_s}$$

M_s: solid-particle mass loaded into the machine (kg)

c_s: solid concentration in the suspension (kg.m^{-3})

$$c_s = \frac{X_s}{\dfrac{X_s}{\rho_s} + \dfrac{1 - X_s}{\rho_L}}$$

X_s: mass fraction of solid matter in the suspension

ρ_s and ρ_L: densities of the solid and the liquid

The basket fill ratio is:

$$P_p = \Omega_{sus} / \Omega_p$$

This rate should be less than 0.5, and Ω_p is the volume of the basket.

2) Filter-cake formation:

The cake forms on the walls of the basket by sedimentation when the suspension is exposed to centrifugal force. The duration of this process is the time that a particle takes to migrate from an initial point on the surface of the suspension to a point on the surface of the final cake.

$$\tau_{FG} = \frac{r_G - r_{L0}}{V_\ell}$$

r_{L0}: radial distance of the initial point on the surface of the suspension (m)

r_G: radial distance of the point on the free surface of the cake (m)

V_ℓ: limiting settling velocity of a particle (m.s^{-1})

The velocity V_ℓ is calculated using equations 3.1.2 and 3.1.5 in Vol. 7, replacing the acceleration due to gravity with the centrifugal force $\omega^2 \overline{r}$

ω: angular rotation velocity of the basket (rad.s^{-1})

$$\omega = 2\pi N$$

N: rotational frequency (rev.s^{-1})

$\overline{r}$: average radial distance for the sedimentation (m)

$$\overline{r} = \frac{1}{2}\left(r_{L0} + r_G\right)$$

where:

$$r_{L0}^2 = r_p^2 - \frac{M_s}{c\pi H} \quad \text{and} \quad r_G^2 = r_p^2 - \frac{M_s}{(1-\varepsilon)\rho_s \pi H}$$

H: (axial) height of the basket (m)

3) Percolation of the clear liquor through the cake:

The volume of the clear liquor is:

$$\Omega_{LC} = H\pi\left(r_G^2 - r_{L0}^2\right)$$

The liquor must pass through the cake and the basket wall. In accordance with Poiseuille's law, we can write:

$$\Delta P = \mu U_L \left(R + \frac{r_p - r_G}{K} \right) = \mu U_L R_{PG}$$ [1.3]

K: permeability of the cake (m^2)

$$K = \frac{d_p^2 \varepsilon^3}{150(1-\varepsilon)^2 (1+\tau)}, \qquad \text{where} \qquad \tau = \frac{1.75 \rho_L U_L d_p}{150 \mu (1-\varepsilon)}$$

τ: turbulent-flow time (generally negligible) (dimensionless)

d_p: size of the solid particles (m)

ε: porosity of the cake

U_L: velocity of the liquid (in an empty bed) through the cake (m.s^{-1})

$$U_L = \frac{dr_L}{d\tau}$$

R: resistance of the filter cloth (m^{-1})

The difference in centrifugal pressure between the free surface of the liquid and the wall of the basket is:

$$\Delta P = \frac{\rho_L \omega^2}{2} \left(r_p^2 - r_L^2 \right)$$

Now:

$$\frac{1}{r_p^2 - r_L^2} = \frac{1}{2 r_p \left(r_L + r_p \right)} + \frac{1}{2 r_p \left(r_p - r_L \right)}$$

Equation [1.1] can be integrated to give the percolation time:

$$\tau_{PE} = \frac{\mu R_{PG} r_p}{\rho_L \omega^2} \left[Ln \left(\frac{r_p + r_G}{r_p + r_{Lo}} \right) - Ln \left(\frac{r_p - r_G}{r_p - r_{Lo}} \right) \right]$$

4) Cake-purge time:

Let Z be the depth of the liquid:

$$Z = r_p - r_L$$

The velocity of the liquid in an empty bed is:

$$U_L = -\varepsilon(1 - S_I)\frac{dZ}{d\tau}$$

ε: porosity of the cake

S_I: irreducible saturation

The braking pressure caused by the cake is:

$$\Delta P_G = \frac{\mu U_L Z}{K} = -\frac{\mu\varepsilon(1 - S_I)}{K}\frac{dZ}{d\tau}Z$$

The braking pressure due to capillarity is per unit area of the basket wall:

$$\Delta P_\gamma = \gamma\cos\theta P$$

γ: surface tension $(N.m^{-1})$

θ: angle of contact

P: wetted perimeter per unit area of the basket wall (m^{-1})

$$P = (1 - \varepsilon)\sigma$$

Hence:

$$\Delta P_\gamma = (1 - \varepsilon)\sigma\gamma\cos\theta$$

θ is the wetting angle.

The centrifugal force exerted on the liquid is:

$$\Delta P_\omega = \rho_L Z(1-\varepsilon)\omega^2 \bar{r}$$

where:

$$\bar{r} = \frac{1}{2}(r_p + r_L) = \frac{1}{2}r_p + \frac{1}{4}(r_p + r_G) = \frac{1}{4}(3r_p + r_G)$$

The braking pressure caused by crossing the wall of the basket is:

$$\Delta P_R = \mu R U = -\varepsilon\mu(1-S_I)R\frac{dZ}{d\tau}$$

The balance of forces is expressed as:

$$\Delta P_G + \Delta P_R = \Delta P_\omega - \Delta P_\gamma$$

That is:

$$-Z\frac{dZ}{d\tau} - RK\frac{dZ}{d\tau} = \frac{K\rho_L Z(1-\varepsilon)\omega^2\bar{r} - (1-\varepsilon)\sigma\gamma K}{\mu\varepsilon(1-S_I)}$$

This Z equation is of the form:

$$-y\frac{dy}{d\tau} - KR\frac{dy}{d\tau} = Ay - B$$

where:

$$\frac{y+KR}{y-B/A} = 1 + \frac{KR+B/A}{y-B/A}$$

$$A = \frac{K\rho_L(1-\varepsilon)\omega^2\bar{r}}{\mu\varepsilon(1-S_I)} \qquad\qquad (\text{m.s}^{-1})$$

$$B/A = \frac{\sigma\gamma\cos\theta}{\rho_L\omega^2\bar{r}} \qquad\qquad (\text{m})$$

After integration (with $Z_0 = r_p - r_G$), the cake purge time is:

$$\tau_{PG} = \frac{Z_0 - Z_1}{A} + \left(\frac{KR + B/A}{A} \right) \ln \left(\frac{Z_0 - B/A}{Z_1 - B/A} \right)$$

Let us agree that:

$$Z_1 = 1.1 \, B/A$$

5) The total centrifuge time is:

$$\tau_{ES} = \tau_{FG} + \tau_{PE} + \tau_{PG}$$

This duration is an upper limit as the three stages are not strictly consecutive and overlap one another. However, we need to add to τ_{ES} a time τ_{CD} for all machine-loading and -unloading operations.

EXAMPLE.–

Case of a spin-dryer processing successive loads of refined sugar. The machine can produce 4200 kg of sugar crystals per hour.

$r_p = 0.6$ m	$\varepsilon_G = 0.38$	$\rho_S = 1500$ kg.m^{-3}
$H = 1$ m	$\mu = 10$ mPa.s	$\rho_L = 1300$ kg.m^{-3}
$N = 2000$ rev.mn^{-1}	$d_p = 0.5 \times 10^{-3}$ m	$\gamma\cos\theta = 0.02$ N.m^{-1}
$R = 109$ m^{-1}	$S_I = 0.11$	$M_S = 190$ kg
	$X_S = 0.25$	

1) Filling the bucket:

$$\Omega_p = \pi \times 0.6^2 \times 1 = 1.13 \text{ m}^3$$

$$c_s = \frac{0.25}{\dfrac{0.25}{1500} + \dfrac{0.75}{1300}} = 336 \text{ kg.m}^{-3}$$

$$\Omega_{sus} = \frac{190}{336} = 0.56 \text{ m}^3$$

We check that $0.56/1.13 = 0.49 < 0.5$

2) Cake formation

$$r_G = \left[0.6^2 - \frac{190}{\pi \times 1 \times 1500(1-0.38)}\right]^{0.5} = 0.543 \text{ m}$$

$$r_{Lo} = \left[0.6^2 - \frac{190}{336\pi \times 1}\right]^{0.5} = 0.424 \text{ m}$$

$$\bar{r} = \frac{1}{2}(0.543 + 0.424) = 0.483 \text{ m}$$

$$\omega^2 \bar{r} = \left(\frac{200 \times 2\pi}{60}\right)^2 \times 0.483 = 209^2 \times 0.483 = 21.186 \text{ m.s}^{-2}$$

$$X = \frac{4 \times 21186 \times \left(0.5 \times 10^{-3}\right)^3 (1500 - 1300)}{3 \times 0.01^2} = 7.062$$

$$Y = \frac{13842}{7.062^{1.97}} + \frac{218}{7.062^{1.074}} = 321.02$$

$$V_{\ell o} = \left[\frac{4 \times 21186 \times 0.01 \times 200}{3 \times 1300 \times 321.02}\right]^{1/3} = 0.513 \text{ m.s}^{-1}$$

$$Re = \frac{0.513 \times 0.5 \times 10^{-3} \times (1500 - 1300)}{0.01} = 5.13$$

$$n = \frac{4.4}{5.13^{0.0982}} = 3.747$$

$$V_\ell = V_{\ell o} \varepsilon_L^n$$

$$\varepsilon_L = \frac{0.75 / 1300}{\dfrac{0.75}{1300} + \dfrac{0.25}{1500}} = 0.776$$

$$V_\ell = 0.513 \times 0.776^{3.747} = 0.198 \text{ m.s}^{-1}$$

$$\tau_{FG} = \frac{0.543 - 0.424}{0.198} = 0.6 \text{ s}$$

3) Percolation of the liquid:

$$K = \frac{\left(0.5 \times 10^{-3}\right) \times 0.38^3}{150 (1 - 0.38)^2} = 2.38.10^{-10} \text{ m}^2$$

$$R_{PG} = 10^9 + \frac{0.6 - 0.543}{2.38 \times 10^{-10}} = 1.239 \times 10^9 \text{ m}^{-1}$$

$$\tau_{PE} = \frac{0.01 \times 1.239 \times 10^{-} \times 0.6}{1300 \times 209^2} \left[\ln \left(\frac{0.6 + 0.543}{0.6 + 0.424} \right) - \ln \left(\frac{0.6 - 0.543}{0.6 - 0.424} \right) \right]$$

$$\tau_{PE} = 0.16 \text{ s}$$

4) Purging of the cake:

$$\bar{r} = \frac{1}{4} (3 \times 0.6 + 0.543) = 0.586 \text{ m}$$

$$A = \frac{2.38 \times 10^{-10} \times 1300 \times 0.62 \times 209^2 \times 0.586}{0.01 \times 0.38 (1 - 0.11)} = 1.452 \text{ m.s}^{-1}$$

$$B/A = \frac{\left(6 / 0.5 \times 10^{-3}\right) \times 0.02}{1300 \times 209^2 \times 0.586} = 7.212 \times 10^{-6} \text{ m}$$

$$KR = 2.38 \times 10^{10} \times 10^9 = 0.238 \text{ m}$$

$$Z_0 = 0.6 - 0.543 = 0.057 \text{ m}$$

$$Z_1 = 1.1 \times 7.212 \times 10^{-6} = 7.933 \times 10^{-6} \text{ m}$$

$$\tau_{PG} = \frac{0.057 - 7.933 \times 10^{-6}}{1.452} + \frac{\left(0.238 + 7.212 \times 10^{-6}\right)}{1.452} \ln\left(\frac{0.057 - 7.212 \times 10^{-6}}{0.721 \times 10^{-6}}\right)$$

$$\tau_{PG} = 1.89 \text{ s}$$

The total spin-drying time is:

$$\tau_{ES} = 0.6 + 0.16 + 1.89 = 2.65 \text{ s}$$

For the duration of loading and unloading operations, let us take:

$$\tau_{CD} = 2 \text{ min.} 40 \text{ s} = 160 \text{ s}$$

Thus, for a single load, the operation will last:

$$160 + 2.65 = 163 \text{ s}$$

The production output will be:

$$P = \frac{190 \times 3600}{163} \# 4200 \text{ kg.h}^{-1}$$

1.4. Deep-bed filtration of liquids

Duroudier [DUR 99] demonstrated that the equation governing the deposition of the particles in a suspension on the grains of a solid is:

$$\frac{\partial(\sigma + \varepsilon)}{\partial t} + \frac{\partial Uc}{\partial z} = 0$$

Let us posit that:

$$\sigma = q(1-\varepsilon) \qquad u = \frac{U}{\varepsilon}$$

u: interstitial velocity of the liquid

ε: porosity of the bed

σ: m^3 of deposited particles per m^3 of bed

q: m^3 of deposited particles per m^3 of solid grains

$$u\frac{\partial c}{\partial z} + \frac{(1-\varepsilon)}{\varepsilon}\frac{\partial q}{\partial t} + \frac{\partial c}{\partial t} = 0$$

This equation is identical to the equation governing adsorption on a fixed bed. It can be solved using similar numerical methods (see sections 2.9 and 2.10 in [DUR 16b]). We shall not go into any further detail here. Duroudier [DUR 99] describes deep beds in detail.

1.5. Filtration of gases

1.5.1. *Aerosols heavily laden with dust (bag filters)*

Such aerosols can be found, in particular, in certain grinding facilities. In this case, the most appropriate filter is the fabric-bag filter.

When the pressure drop reaches about 2500 Pa, the cake is mechanically removed and is dropped into a funnel located beneath the filter.

The law governing the operation of the filter is:

$$\Delta P = \mu V (R + M_s \alpha)$$

M_s: mass of solid per unit area $(kg.m^{-2})$

α: specific resistance $(m.kg^{-1})$

V: face velocity of the gas (in an empty bed) $(m.s^{-1})$

The time between two filter-unclogging operations is a few minutes, and is expressed as:

$$\tau_{dcom} = \frac{M_{s\,max}}{Vc}$$

c: dust content of the gas (kg.m^{-3})

The maximum mass of dust per unit area is of the order of 0.5 kg.m^{-2}.

An example is provided in [DUR 99].

1.5.2. *High-efficiency filters (HEFs)*

The gas – generally air – contains very little dust, but it needs to be completely dust-free. This is a requirement in "clean rooms", such as high-precision-electronics workshops or pharmaceutical laboratories.

Instead of a filter cloth, the device used is a highly porous fiber mat, which produces a minimum pressure drop.

The pressure drop is given by Lamb's formula [LAM 63], complemented by Pich [PIC 87].

$$\frac{\Delta P}{L} = \frac{32\beta\mu V}{(1 + \ln\beta)d_{eq}}$$

L: mat thickness

β: ones' complement of the porosity ($\beta = 1 - \varepsilon$)

If there are any dust particles trapped in the fibers:

$$\beta = \beta_{fibers} + \beta_{dust}$$

If the dust particles are in the shape of fibers of diameter d_{PF}:

$$\frac{\beta}{d_{eq}} = \frac{\beta_{fibers}}{d_{fibers}} + \frac{\beta_{dust}}{d_{dust}}$$

If the dust particles are spheroidal in shape, we write:

$$\frac{4\beta}{d_{eq}} = \frac{4\beta_{fibers}}{d_{fibers}} + \frac{6\beta_{dust}}{d_{dust}}$$

However, the fibers may be compressed together, as in felt, or even paper or cardboard (air filters for automobiles). In these latter embodiments, the porosity is markedly reduced, and we need to use Ergun's formula [ERG 52] for the pressure drop, rather than Lamb's [LAM 63].

The collection efficiency may be calculated using the equation formulated by Stechkina et al. [STE 69]. An example is investigated by [DUR 99] for a high-efficiency filter and a very-high-efficiency filter.

1.6. Evacuation of the liquid by compression

Shirato et al. [SHI 79] studied this phenomenon with a constant, uniaxial pressure, and provided a simplified equation to express the consolidation U of the cake.

$$U = \frac{L_o - L}{L_o - L_\infty}$$

Shirato et al. [SHI 86] used the Voigt model (spring + viscous damper) to express the consolidation.

The entire compression process is discussed in Chapter 7 of this volume.

1.7. Screening

1.7.1. General considerations

We have seen how to separate a fluid and a divided solid. Let us now discuss how to separate, within a divided solid, (i) the particles whose size is greater than a given diameter, and (ii) the particles whose size is less than that diameter. To do so, the particles are placed on a vibrating surface with apertures calibrated to allow the passage of small particles. The large particles will not pass through and will remain on the surface.

Today, this operation is performed on an industrial scale, in a continuous process, using devices called screens.

1.7.2. Screens

1) Static devices

Let us call to mind, as a reminder, the conventional riddle, which comprises a frame on which a sturdy, wire mesh, inclined at a 45-degree angle, is held in place with a stay (work-site screen for sand, gravel, etc.).

To remove small particles prior to crushing, rigid-bar screens, such as grizzlies, are used. The distance between the bars of the screen varies from 2 to 20 cm; they are inclined at an angle of 20° to 30°, and the bars offer a high degree of mechanical resistance (manganese steel).

Wedge-wire screens are screens in which the wires of the mesh are set very close together (with a spacing ranging from a few tenths of a millimeter to 2 to 3 mm. They have long been used as drip grates.

2) Slow-moving devices

The screening occurs between two, stacked disk rollers. These devices are suitable for coarse feeds.

3) Rotating screens (trommels)

These screens comprise a metallic cylinder perforated along its length with 2 or 3 series of holes which increase in size as they approach the output end. The trommel rotates about its axis, at a slope of 10 to 20%. The rotation speed is low (20–30 rev.mn^{-1}). The diameter may be as much as 2m, and the length as much as 10m. This type of device is used either dry – to screen sable and/or gravel – or with water, in which case very small particles are removed by washing or de-sludging (sand quarries, dredgers, etc.).

Flour-sifting devices comprise a very fine cloth mounted on a frame, the cross-section of which may be either hexagonal or circular.

4) Shaking screens

The screen is suspended from the ceiling by steel springs or blades, or attached to the floor by elastic rods inclined from the horizontal. A connecting rod forms a link between the screen and an eccentric crank.

Resonance screens actually comprise two screens on the same frame, oscillating in phase opposition with each other and coupled by elastic elements such that the period of oscillation of the whole arrangement is the desired operating period. Theoretically, the work required to maintain the oscillations only has to make up for the energy lost due to damping.

5) Vibrating screens

Vibrating screens are becoming more popular on the market, to the detriment of other types of devices. They are described in this next section.

1.7.3. *Vibrating screens*

Vibrating screens are the most commonly used devices, because their design is simple and they lend themselves well to a very wide variety of applications. The material advances due to a combination of the slope and the circular motion given to the riddle.

That circular motion is created by an unbalanced electrical motor for small devices or by an unbalanced shaft vibrator (so-called double-deck screen) or indeed by a mechanical vibrator containing a shaft and two eccentric bearings (so-called quadruple bearing screen).

Quadruple-deck screens (or eccentric screens) have the advantage of an invariable amplitude of movement and, in particular, independent of the load on the screen. The operation of unbalanced screens, on the other hand, may be "stifled" by the action of an accidental overload if they are too lightly designed.

Depending on the material being treated, the slope may vary from 14 to 24°. The rate of progression of the material on the screen can vary from 0.2 to 0.8 m.s^{-1}.

The breadth ranges between 0.5 and 2.5 m, with lengths of between 1 and 7.2 m. The maximum feed flowrate must be compatible with the breadth of the screens. They may be equipped with 1–4 superposed mats. The attack speeds vary from 650 to 3000 rev.mn^{-1} for courses (diameter of the circular trajectory) dropping from 12 to 3 mm. The necessary powers of the motors range from 1 to 35 kW.

These screens are sometimes operated on a counter-current basis when the efficiency of the screening needs to win out over production.

Screens with several superposed mats simplify the installation, but it is always to the detriment of the efficiency. In any case, the ratio of the mesh apertures of two successive mats must be no less than 1/3. It is often equal to $1/\sqrt{2}$.

The usual field of application is dry screening, between 0.1 mm and 120 mm. With wet screening, it is possible to go down to 30 μm.

1.7.4. *Screening surface*

The screening surface is generally composed of a metal cloth. The mesh is generally square, and the ratio of the surface available for the passage to the surface of the cloth – i.e. the degree of aperture – is:

$$F_0 = \left(\frac{p}{p+f} \right)^2$$

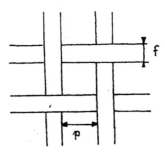

Figure 1.2. *Square mesh*

If the mesh is rectangular, with sides measuring p_1 and p_2, with the wires being f_1 and f_2 in diameter, the degree of aperture is:

$$F_0 = \frac{p_1 p_2}{(p_1 + f_1)(p_2 + f_2)}$$

In the case of splits whose width is p, separated by metal over a breadth f:

$$F_0 = \frac{p}{p+f}$$

From the point of view of screening, the "aperture diameter" is the side of a square mesh exhibiting the same performances:

$$d_0 = F_e p \cos \alpha$$

p is defined as above (Figure 1.2)

α is the angle of the screening surface in relation to the horizontal

F_e is the equivalence coefficient. It is given by the table below

Length/width ratio	F_e
L/1 < 2	1
2 < L/1 < 4	1.1
4 < L/1 < 25	1.2
25 < L/1	1.4

The table in Appendix 1, for square meshes, gives the apertures and wire diameters used in France.

1.7.5. Characteristics of the feed to a vibrating screen

From the standpoint of the operation of screening, the feed is defined by:

– its histogram in terms of PSD. A laboratory screening test on a sample of mass M indicates the mass m_i between the dimensions d_{i-1} and d_i. That slice is characterized by the mean dimension:

$$\overline{d_i} = \sqrt{d_{i-i}.d_i}$$

With this dimension, we associate the fraction:

$$Z_i = m_i / M$$

– the screenable or passthrough, which is the mass fraction of the feed whose size is less than the mesh aperture of the screen and which should, theoretically, pass through the mat of the screen;

– the theoretical retention, which is the complement of the previous fraction.

On exiting the screen, therefore, the feed is divided into:

– the passthrough which has, in fact, come through the mat;

– the refuse, which has not passed through the mat.

Fine grains, which are significantly smaller than the aperture, pass through with no difficulty. On the other hand, those whose size begins to approach that of the aperture have decreasing chances of passing through. Thus, the probability of passage of grains of $0.9 d_0$ is 1%, meaning that the grain should encounter 100 apertures on its path to pass through.

In quantitative terms, the fineness of the screening is characterized by the half-aperture fraction, meaning the weight fraction of grains whose size is less than half of the aperture. We shall call this fraction "the fine fraction".

Grains larger than 1.25 times the aperture have no chance of passing through, even if they are elongated in form. Thus, they remain on the mat and "dilute" the screenable grains and hamper their approach to the screening surface. This holds true between 1 and 1.25 times the aperture, but certain grains do pass through. We characterize the presence of all these grains by the theoretical refuse in the feed whose size is greater than the aperture.

Furthermore, we apply the term "difficult grains" to those whose size is between 0.75 and 1.25 times the aperture.

The degree of humidity in relation to the total wet flowrate may be an important factor. Damp fibers tend to clog the mat. The screening capacity gradually falls to zero as the humidity approaches 20%. On the other hand, if we add the same mass of water to the solid, we see that the screen is restored, but in varying proportions depending on the aperture.

The form of the grains also comes into play. If they are round (as in the case of grains of sand), they pass through easily. The difficulty increases if the shape of the grains approaches an angular form with concavities of the surface, or indeed a plate-like or needle-like shape. In the latter case, the passage surface will be constituted by slits.

The density of the solid also comes into play. The porosity ε of the loose solid varies between 0.4 and 0.5, and the apparent density is:

$$\rho_a = (1-\varepsilon)\rho_s$$

where ρ_s is the true density of the solid. The denser the grains, the greater the effect of gravity.

All the grains have three dimensions:

– minimum;

– intermediary;

– maximum.

It is the intermediary dimension which defines the screenable nature.

Finally, the feed will be characterized by:

– a flowrate in tonnes per hour;

– the percentage of fine particles Z_F;

– the percentage of theoretical refuse Z_R;

– the percentage of difficult grains Z_D;

– the percentage humidity in relation to the total flowrate;

– the shape of the grains;

– the apparent density (loose).

1.7.6. *Working parameters and manufacture parameters*

The slope increases the flowrate but decreases the surface of passage of the grains.

The influence of the frequency and amplitude of the vibrations has not been studied. The finer the grains in the feed, the higher the frequencies used.

The aperture of the mesh is the crucial parameter. In section 1.7.4, we give the fraction of surface available for passage, corresponding to each aperture.

When several superposed mats are present, we can assume that the passthrough from the top mat is no greater than 90% of the surface of the mat situated immediately below it.

Width (mm)	Flowrate (m^3/h)
500	50
750	90
1000	135
1250	190
1500	250
1750	320
2000	400
2500	600

Table 1.1. *Maximal flowrate of the screen depending on its width*

1.7.7. *Screen-performance criteria*

The performances of a screen are characterized by:

1) The refuse ratio for each slice of grains:

$$r_i = \frac{\text{refuse for size i}}{\text{screenable for size i}}$$

2) The overall passthrough yield for the chosen flowrate:

$$\eta = \frac{\text{passthrough}}{\text{screenable}} = \sum (1 - r_i) z_i$$

1.7.8. *Aforementioned considerations in quantitative terms*

The first serious effort was made by Bauman and Ermolaev [BAU 70]. Karra [KAR 79] went on to improve the method. It is on these works that our discussions will be based.

1) The standard passthrough flowrate per unit surface area is obtained by the relation:

$$\varnothing^0 = \varnothing_M F_F F_R F_D F_H F_G F_p F_S F_P F_A \left(\text{tonne.h}^{-1}.\text{m}^{-2} \right)$$

$\varnothing_M$ is the base flowrate, which is a function of the aperture used:

$$\varnothing_M = 12.129 \, d_0^{0.3162} - 10.299 \qquad\qquad d_0 < 50.8 \text{ mm}$$

$$\varnothing_M = 0.339 \, d_0 + 14.412 \qquad\qquad d_0 \geq 50.8 \text{ mm}$$

F_F expresses the presence of fine particles:

$$F_F = 0.012 Z_F + 0.7 \qquad\qquad Z_F \leq 30\%$$

$$F_F = 0.153 Z_F^{0.564} \qquad\qquad 30\% < Z_F < 55\%$$

$$F_F = 0.0061 Z_F^{1.37} \qquad\qquad 55\% < Z_F < 80\%$$

$$F_F = 0.05 Z_F - 1.5 \qquad\qquad Z_F \geq 80\%$$

F_R is a function of the theoretical refuse in the field:

$$F_R = -0.012 \, Z_R + 1.6 \qquad\qquad Z_R \leq 87\%$$

$$F_R = 0.0425 \, Z_R + 4.275 \qquad\qquad Z_R > 87\%$$

F_D expresses the influence of the difficult grains:

$$F_D = 0.844\left(1 - \frac{Z_F}{100}\right)^{3.453}$$

F_H is a function of the humidity X of the feed:

F_H	1	0.8	0.5	0.4	0.3	0.2	0
X	$\leq 4\%$	5%	6%	8%	9%	10%	20%

F_G is the form function of the grains:

Form	F_G
Round particles (sand)	1
Stone or ground ore	0.85
Highly irregular	0.70

F_M is the apparent density function of the grains:

$$F_\rho = \frac{\rho_a}{16000} \qquad \left(\rho_a \text{ in kg.m}^{-3}\right)$$

F_s expresses the superposition of mats:

Mat n°	F_s
1	1
2	0.9
3	0.8
4	0.7

F_p varies with the slope of the mat:

Slope	F_p
20%	1
Horizontal	0.8

F_A comes into play for a watered screen, and depends on d_0.

Aperture (mm)	F_A
5	3.5
4 or 8	3
3 or 10	2.5
2 or 15	1.75
20	1.5
30	1.25
– by 1 or + by 30	1

2) The overall pass yield η is determined as follows:

$$\eta = 100 \left(\frac{\varnothing^0}{\varnothing} \right)^{0.148}$$

$\varnothing$ is the theoretical passthrough (tons.m^{-2}) corresponding to the feed:

$$\varnothing = \frac{W}{A} \left(1 - \frac{Z_R}{100} \right)$$

W: feed flowrate (t/h)

A: surface area of the mat (m²)

3) PSD of the refuse

According to Karra [KAR 79], the cutoff diameter d_{50} of the refuse ratio is deduced simply from the yield by:

$$d_{50} = \eta d_0$$

The refuse ratio for grains of size d_i is:

$$r_i = 1 - \exp\left[-0.693 \left(\frac{d_i}{d_{50}} \right)^{5.846} \right]$$

From this, for example, we can deduce the passthrough flowrate for the size d_i.

$$W_{pi} = WZ_i \left(1 - r_i \right)$$

EXAMPLE.–

Consider that we want to treat an ore with the aperture 2 mm on a screen inclined 20° from the horizontal. The feed flowrate is 25 tonne.h^{-1} for a clear area of 1.5 m × 3.7 m. The screen contains only a mat.

The PSD of the feed is such that:

$$Z_F = 48\% \qquad\qquad Z_R = 22\% \qquad\qquad Z_D = 30\%$$

The screen works dry.

The apparent density is 1200 kg m^{-3}

The application of the above formulae gives:

$$\varnothing_M \left(d_0 \right) = 12.129 \times 2^{0.3162} - 10.299 = 4.8 \text{ tonne.h}^{-1}.\text{m}^{-2}$$

$$F_F = 0.153 \times 48^{0.564} = 1.358$$

$$F_R = -0.012 \times 22 + 1.6 = 1.864$$

$$F_D = 0.844(1 - 0.3)^{3.453} = 0.246$$

$$F_M = 1$$

$$F_G = 0.85$$

$$F_\varphi = \frac{1200}{1600} = 0.75$$

$$F_S = 1$$

$$F_P = 1$$

$$F_A = 1$$

Hence:

$$\varnothing^0 = 4.8 \times 1.358 \times 1.864 \times 0.246 \times 1 \times 0.85 \times 0.75 \times 1 \times 1 \times 1$$

$$\varnothing^0 = 1.91 \text{ tonne.h}^{-1}.\text{m}^{-2}$$

The theoretical passthrough is

$$\varnothing = \frac{25}{1.5 \times 3.7}(1 - 0.22) = 3.51 \text{ tonne.h}^{-1}.\text{m}^{-2}$$

The overall pass yield is:

$$\eta = 100\left(\frac{1.91}{3.51}\right)^{0.148}$$

$$\eta = 91\%$$

$$d_{50} = 0.91 \times 2 = 1.82 \text{ mm}$$

The refuse for the diameter d_i is:

$$r_i = 1 - \exp\left[-0.693\left(\frac{d_i}{1.82}\right)^{5.846}\right]$$

In conclusion, this result could be compared with that found by Karra [KAR 79] method.

Theory of Membrane Filtration

2.1. Introduction

2.1.1. *Convective filtration: definition*

Conventional filtration conducted on fine particles whose size is less than 1μm would immediately give rise to the formation of a cake which soon becomes impermeable to the filtrate. We can remedy this difficulty by circulating the prefilter solution along the filtration support.

Thus, the cake is entrained as soon as it forms.

In the same way as convective heat transfer takes place across a wall along which the fluid circulates, convective filtration takes place across a membrane along which the prefilter solution circulates. One might hesitate to use the term "convective filtration", because the adjective "convective" qualifies a term in the supplemented Nernst–Planck equation (see section 2.3.6).

This technique is used not only for microparticles, but also for reverse osmosis, the aim of which is to filter molecules. The filtration cake to be evacuated is then replaced by the polarization layer, whose thickness decreases with the sweeping rate.

Note that convective filtration can be used for particles which are much larger than 1 μm. Thus, the company Alfa-Laval has patented a filter using a rotating drum around a horizontal axis whose rotation speed is around 20 times greater than the conventional speed of such a filter. The cake then no longer accumulates on the filter cloth, and is entrained by the friction of the prefilter solution, which is thus concentrated into particles. Naturally, this filter is primarily a suspension thickener.

In practice, filtration is performed in different ways depending on whether:

– the membrane is electrically inert;

– the membrane is an ion exchanger.

Maurel [MAU 77] reviews the processes of membrane-based separation. Similarly, Applegate [APP 84] describes the principle of these processes.

On the basis of the size of the particles, we can construct Table 2.1 of membrane-based filtration operations when the membrane is electrically inert.

Rumeau [RUM 85] reviews the various types of filtration (capillary flow, cutoff zone, and transfer by solubilization/diffusion). Michaelis *et al.* [MIC 65] studied transport across a membrane by reverse osmosis. Diffusion is predominant.

Diameter of the particles	*Type of prefilter solution and concentration mode*
$d_p < 0.5$ nm	reverse osmosis (or hyperfiltration)
0.5 nm $< d_p < 5$ nm	nanofiltration
5 nm $< d_p < 200$ nm	colloids (sols) (ultrafiltration)
200 nm $< d_p < 1$ μm	micro-fine particles (microfiltration)
1 μm $< d_p < 20$ μm	very fine particles (microfiltration)
20 μm $< d_p < 100$ μm	fine particles (centrifugation or conventional filtration)
$d_p > 100$ μm	gravity-based decantation

Table 2.1. *Membrane filtration*

Ion-exchange membranes are made up of groups generally of the same nature but *all of the same sign*, affixed to the skeleton of the polymer forming the membrane. The *free ions* circulating through the skeleton of the membrane are:

– either of the same sign as the fixed groups. These are co-ions, which we shall call "c-ions" for short. *These ions are very infrequent*;

– or of the opposite sign to that of the fixed groups. These are counter-ions, known as "g-ions" (from the German *gegen*, meaning "counter").

A membrane is said to be positive and cationic if its g-ions are positive, and to be negative and anionic if its g-ions are negative.

Obviously, only the g-ions can be exchanged.

2.2. Single solute and electrically inert membrane

2.2.1. Definitions

Figure 2.1 shows the cross-section of a membrane in convective filtration, and the variation of the concentration across the thickness of the membrane.

The intrinsic work of the membrane is expressed by the concept of transmission Tr_i.

$$Tr_i = \frac{c_P}{c_E}$$

c_E and c_P: inlet- and outlet concentrations (in the filtrate) (kmol.m^{-3})

However, a very commonly used notion is that of the rejection R, which is the ones' complement of the transmission.

$$R + Tr = 1, \text{ meaning that } R = \frac{c_E - c_P}{c_E}$$

Let us set:

J_s and J_e: flux density of the solute and the solvent (water) (kmol.m^{-2}.s^{-1})

J_v: volumetric flux density (m^3.m^{-2}.s^{-1} = m.s^{-1})

$$J_v = V_s J_s + V_e J_e$$

The parameter J_v is imposed by the external pumping energy.

V_s and V_e: molar volumes of the solute and of water (m^3.kmol^{-1})

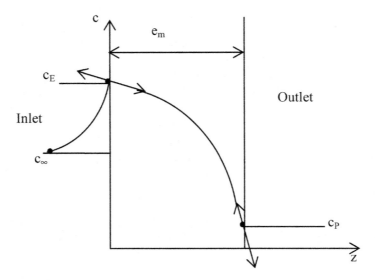

Figure 2.1. *Concentration in the membrane*

For an electrically neutral solute and at the surface of the membrane, the concentration on the outer side of the membrane is equal to the concentration on the inner side. In the membrane, *the concentrations are evaluated in relation to the impregnating liquid phase*.

2.2.2. *Polarization layer, and observable- and intrinsic transmissions*

The equation of the polarization layer is obtained by writing that the solute brought by the liquid stream toward the membrane is evacuated by diffusion.

$$J_v\left(c - c_p\right) - D\frac{dc}{dz} = 0$$

This equation, which contains a convection term and a diffusion term, can immediately be integrated.

$$\frac{c_E - c_P}{c_\infty - c_P} = \exp\left(\frac{J_v \delta_P}{D}\right) = \exp\left(\frac{J_v}{\beta}\right) \qquad [2.1]$$

c_∞: concentration of the prefilter solution (kmol.m^{-3})

c_E: concentration at the inlet of the membrane (kmol.m^{-3})

β: coefficient of re-entrainment of the solute. That coefficient is a conventional material-transfer coefficient

$$\beta = \frac{D}{\delta_P}$$

D: diffusivity of the solute (m^2.s^{-1})

δ_P: thickness of the polarization layer (m)

Let us divide the top and bottom of the left-hand side of equation [2.1] by c_P. We obtain:

$$\frac{c_E}{c_P} = 1 + \left(\frac{c_\infty}{c_P} - 1 \right) \exp\left(\frac{J_V}{\beta} \right)$$

However, the transmissions are:

$$Tr_o = \frac{c_P}{c_\infty} (\text{observable}) \quad \text{and} \quad Tr_i = \frac{c_P}{c_E} (\text{intrinsic}) \qquad [2.2]$$

Thus:

$$Tr_i = \frac{1}{1 + \left(\dfrac{c_\infty}{c_P} - 1 \right) \exp\left(\dfrac{J_V}{\beta} \right)} = \frac{\dfrac{c_P}{c_\infty}}{\dfrac{c_P}{c_\infty} + \left(1 - \dfrac{c_P}{c_\infty} \right) \exp\left(\dfrac{J_V}{\beta} \right)}$$

or indeed:

$$Tr_i = \frac{Tr_o}{Tr_o + (1 - Tr_o) \exp\left(\dfrac{J_V}{\beta} \right)} \qquad [2.3]$$

NOTE.–

Using equation [2.1], it is possible to obtain an expression of the optimal concentration of the feed for a minimum duration of an operation of diafiltration. This operation consists of circulating a solution, passing it through a membrane.

2.2.3. *Thickness of the polarization layer*

To begin with, it is useful to determine the thickness δ of the polarization layer. We write:

$$\delta = D / \beta$$

D: diffusivity of the solute $(m^2.s^{-1})$

β: transfer coefficient $(m.s^{-1})$

Except in microfibers, the flow very generally takes place in the turbulent regime. However, we shall give both expressions of the Sherwood number.

$$Sh = \beta \, D_h/D$$

D_h: hydraulic diameter of the flow pipe (m)

$$D_h = 4A/P$$

A: section of the pipe (m^2)

P: wetted perimeter (giving rise to transfer) (m)

Nature of the flow	Sherwood number
Laminar (Lévêque)	$Sh = 1.62 \, (Re \, Sc \, D_h/L)^{1/3}$
Turbulent (Geissler)	$Sh = 0.023 Re^{0.8} Sc^{0.33}$

Table 2.2

L: length of the pipe (m)

Re: Reynolds number

Sc: Schmidt number

$$Re = \frac{4W}{\mu P} = \frac{VD_h \rho}{\mu} \quad Sc = \frac{\mu}{\rho D}$$

W: mass flowrate $(kg.s^{-1})$

V: flow velocity $(m.s^{-1})$

μ: viscosity (Pa.s)

ρ: density $(kg.m^{-3})$

A quick calculation shows that δ increases as $D^{2/3}$ does.

EXAMPLE.–

$D_h = 0.025$ m $D = 10^{-9}$ m².s⁻¹ $\mu = 10^{-3}$Pa.s

$V = 1$ m.s⁻¹ $\rho = 1000$ kg.m⁻³

$$Re = \frac{1 \times 0.025 \times 1000}{10^{-3}} = 25000$$

$$Sc = \frac{10^{-3}}{1000 \times 10^{-9}} = 1000$$

$$Sh = 0.023 \times 25000^{0.8} \times 1000^{0.33} = 758.5$$

$$\delta = \frac{D}{\beta} = \frac{D_h}{Sh} = \frac{0.025}{758.5} = 3.29.10^{-5}\,m$$

NOTE.–

Sherwood *et al.* [SHE 65] conducted a very in-depth study of the polarization layer.

2.2.4. *Case of large molecules, macromolecules, microparticles*

Geissler's relation, for these "solutes", leads to too low a degree of polarization. Rautenbach *et al.* [RAU 88] propose the following relations for calculating the Sherwood number.

Ultrafiltration $\qquad\qquad\qquad$ $Sh = 4 \times 10^{-2} \, Re^{0.8} Sc^{0.4}$

Microfiltration $\qquad\qquad\qquad$ $Sh = 4.7 \times 10^{-5} \, Re^{1.26} Sc$

The diffusivity is calculated by the Stokes–Einstein formula:

$$D = \frac{kT}{6\pi\mu r}$$

k: Boltzmann's constant (1.38048×10^{-23} J.K^{-1})

μ: viscosity of the liquid (Pa.s)

r: radius of the molecule or of the particle (m)

This radius can be obtained on the basis of the molar or particle volume V:

$$\frac{4\pi r^3}{3} = V = \frac{M}{\rho}$$

M: molar mass (kg.kmole^{-1} or kg.kiloparticle^{-1})

ρ: density in the liquid state (or solid state) (kg.m^{-3})

2.2.5. *Intrinsic transmission (in the membrane)*

The elementary equation of transport *in the membrane* is written:

$$J_s = K_c J_V c - K_d D_\infty \frac{dc}{dz} = \text{convection} + \text{diffusion} \qquad\qquad [2.4]$$

The coefficients K_c and K_d are, respectively, the steric attenuation coefficients for convection and diffusion. They are dimensionless and they characterize the *motion* of the solute in the pores.

J_s: molar flux density of the solute ($kmol.m^{-2}.s^{-1}$)

J_V: permeation flux in volume per unit surface of the membrane ($m^3.m^{-2}.s^{-1}$, which is $m.s^{-1}$)

z: path taken in the membrane from the inlet (m)

D_∞: diffusivity of the solute at infinite dilution in the solvent ($m^2.s^{-1}$)

The coefficients K_c and K_d are given by Bandini and Vezzani [BAN 03]. We set:

$$\lambda = \frac{\text{radius of the solute}}{\text{mean pore radius}}$$

For K_c, which is the coefficient for convection:

$0 < \lambda \leq 0.8$ $\qquad\qquad K_c = 1 + 0.054\lambda - 0.988\lambda^2 + 0.441\lambda^3$

$0.8 < \lambda \leq 1$ $\qquad\qquad K_c = -6.830 + 19.348\lambda - 12.518\lambda^2$

For K_d, which is the coefficient for diffusion:

$0 < \lambda < 0.8$ $\qquad\qquad K_d = 1 - 2.30\lambda + 1.154\lambda^2 + 0.224\lambda^3$

$0.8 < \lambda \leq 1$ $\qquad\qquad K_d = -0 \times 105 + 0.318\lambda - 0.213\lambda^2$

The transport equation [2.4] can easily be integrated over the thickness of the membrane e_m.

$$\frac{e_m J_V K_c}{D_\infty K_d} = Pé = Ln\left[\frac{c_P - J_s / K_c J_V}{c_E - J_s / K_c J_V}\right] \qquad [2.5]$$

However:

$$\frac{J_s}{J_V} = c_P$$

We shall set:

c: quantity of solute per unit volume of the solution

c_m: quantity of the same solute per unit volume of the impregnated membrane

We have:

$$\frac{c_m}{c} = \frac{\text{volume of liquid}}{\text{total volume of the membrane}} = \varepsilon$$

ε: porosity of the wet membrane

ε is also the coefficient of sharing between the solution and the membrane.

Pé: Péclet number (ratio of the convective transport to the diffusive transport, dimensionless)

Equation [2.5] becomes:

$$Tr_i = \frac{c_{mP}}{c_{mE}} = \frac{\varepsilon K_c \exp Pé}{\varepsilon K_c - 1 + \exp Pé} \qquad [2.6]$$

We can see that:

– if the permeation flowrate increases indefinitely – i.e. if Pé tends toward infinity:

Tr_i tends toward $\varepsilon K_c < 1$

– if the permeation flowrate J_V becomes very low and tends toward zero, it is then advisable to bring into play the rejection:

$$R_i = 1 - Tr_i = \frac{\varepsilon K_c - 1 + \exp Pé - \varepsilon K_c \exp Pé}{\varepsilon K_c - 1 + \exp Pé} = \frac{(1 - \varepsilon K_c)(\exp Pé - 1)}{\varepsilon K_c - 1 + \exp Pé}$$

If Pé tends toward zero:

$\exp Pé \sim 1 + Pé$

We then obtain:

$$R_i \sim Pé\frac{(1-\varepsilon K_c)}{\varepsilon K_c} = \frac{e_m(1-\varepsilon K_c)}{D_\infty \varepsilon K_d}J_v$$

Hence:

$$R_i = \frac{J_v}{D_\infty/e_m}\left(\frac{1-\varepsilon K_c}{\varepsilon K_d}\right)$$

2.2.6. Variations in the observable transmission

By equaling the two expressions [2.3] and [2.6] of Tr_i, we obtain an expression for the observable transmission.

$$Tr_o = \frac{\varepsilon K_c}{(\varepsilon K_c - 1)\exp\left[-\left(Pé + \dfrac{J_v}{\beta}\right)\right] + (1-\varepsilon K_c)\exp\left(\dfrac{-J_v}{\beta}\right) + \varepsilon K_c}$$

This function exhibits a minimum which is obtained by zeroing the derivative of the denominator in relation to J_v.

$$(\varepsilon K_c - 1)\left(\frac{\varepsilon K_c e_m}{\varepsilon K_d D_\infty} + \frac{1}{\beta}\right)\exp-\left(Pé + \frac{J_v}{\beta}\right) + \frac{1}{k}(1-\varepsilon K_c)\exp\left(-\frac{J_v}{\beta}\right) = 0$$

This means that:

$$\exp Pé = \frac{K_c \beta e_m}{K_d D_\infty} + 1$$

$$Pé = \frac{K_c e_m J_v}{k_d D_\infty} = Ln\left(1 + \frac{K_c e_m \beta}{K_d D_\infty}\right)$$

Thus, we obtain the value of J_v corresponding to the minimum of the observed transmission.

Opong and Zydney [OPO 91] verified the above calculations experimentally. The shape of the variations of Tr_o as a function of J_v is given by Figure 2.2.

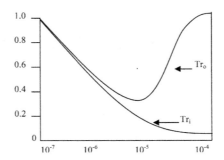

Figure 2.2. *Shape of the variations of the transmissions as a function of J_V $(m.s^{-1})$*

NOTE.–

If we accept that the flow of the solvent in the pores obeys Poiseuille's law, we can write:

$$J_V = \frac{r_p^2 \Delta P}{8 \mu e_m}$$

The Péclet number then becomes:

$$Pé = \frac{r_p^2 \Delta P}{8 \mu D_\infty} \frac{K_e}{K_d}$$

We see that *the Péclet number has become independent of the thickness of the membrane.*

However, we can only accept the validity of Poiseuille's law if the diameter of the pore is at least equal to ten times the molecular diameter of the solvent. Thus, for water, this condition is expressed by:

$$d_{pore} \geq 10 \times 0.3 = 3 \text{ nm}$$

For nanofiltration membranes:

$$0.5 \text{ nm} < d_{pore} < 5 \text{ nm}$$

Nevertheless, even for these membranes, the flux density of permeation remains directly proportional to ΔP, but the pore radius r_p which plays a part in the coefficient of proportionality is not necessarily the true radius of the pores.

2.2.7. Apparition of a gel

According to equation [2.1], on contact with the membrane the concentration c_c increases exponentially with J_V:

$$c_E = c_P + (c - c_P) \exp\left(\frac{J_V}{\beta}\right)$$

However, the concentration c_c cannot increase indefinitely because, for geometric reasons, the volume fraction occupied by the solute can be no greater than a value of the order of 0.5 or 0.6. Thus, we have:

$$c_c V_s < 0,6$$

V_s: volume of a kilomole or of a kiloparticle ($m^3.kmol^{-1}$ or $m^3.kpart$)

The return flux of the solute (from the membrane to the feed solution) by back diffusion Ddc/dz (see equation [2.4]) is thus limited, and we see the accumulation of solute against the membrane, but always at the same concentration $c_G = c_E$.

In other words, an additional hydraulic resistance is established between the prefilter solution and the membrane. This resistance is that of a gel layer whose thickness is δ_G and whose permeability is K_G. The pressure drop across the gel is therefore:

$$\Delta P_G = \frac{\mu J_V \delta_G}{K_G}$$

Remember that:

$$K_G = \frac{\varepsilon_G^3 d^2}{150(1 - \varepsilon_G)^2} \qquad \text{[DUR 99]}$$

ε_G: porosity of the gel (liquid fraction in terms of volume)

d: diameter of a particle (or of a macromolecule) of the gel (m)

Thus, if we increase the driving pressure, the thickness δ_G of the gel increases proportionally, and the flux density J_V of the filtrate remains constant.

2.2.8. The Maxwell–Stefan equation and dialysis (see discussion in section 4.2, Vol. 1)

For the solute and in the membrane, the M–S equation is written,

for the solute:

$$-RT\Gamma_s \frac{dc_s}{dz} - \phi_s \frac{dP}{dz} = \left(x_e J_s - x_s J_e\right)\frac{RT}{D} + RT\frac{J_s}{D_{ms}}$$

for the water:

$$-RT\Gamma_e \frac{dc_e}{dz} - \phi_e \frac{dP}{dz} = \left(x_s J_e - x_e J_s\right)\frac{RT}{D} + RT\frac{J_e}{D_{me}}$$

$$\Gamma_s = 1 + \frac{d \operatorname{Ln}\gamma_s}{d \operatorname{Lnc}_s} \qquad\qquad \Gamma_e = 1 + \frac{d \operatorname{Ln}\gamma_e}{d \operatorname{Lnc}_e}$$

γ_s and γ_e: activity coefficients of the solute and of water

ϕ_s and ϕ_e: volumetric fractions of the water and the solute in the liquid

D: reciprocal diffusivity of the solute and the water for a given concentration (see section 4.6.2 in Vol. 1) $(m^2.s^{-1})$

D_{ms} and D_{me}: individual diffusivity of the solute and water in the membrane (m^2)

The pressure equation is as follows (because $\phi_s + \phi_e = 1$):

$$\frac{dP}{dz} = -RT\left(\frac{J_e}{D_{me}} + \frac{J_s}{D_{ms}}\right)$$

In addition:

$$c_e V_e + c_s V_s = 1 \quad \text{so} \quad c_e = \frac{1 - c_s V_s}{V_e}$$

Finally, let us set:

$$J_e = \frac{J_s}{\kappa} \qquad x_e = \frac{c_e}{c_T} \quad \text{and} \quad x_s = \frac{c_s}{c_T}$$

In the M-S equation of the solute, let us replace dP/dz with its expression:

$$\Gamma_s \frac{dc_s}{dz} = c_s \left(\frac{V_s J_s}{D_{me} \kappa} + \frac{V_s J_s}{D_{ms}} + \frac{V_s J_s}{c_T V_e D} - \frac{J_s}{c_T \kappa D} \right) - J_s \left(\frac{1}{D_{ms}} + \frac{1}{c_T V_e D} \right)$$

Therefore, *if we accept that $\Gamma_s = 1$ and $\Gamma_e = 1$:*

$$\frac{1}{\Sigma J_s} \frac{dc_s}{dz} = c_s - \frac{\Delta}{\Sigma}$$

By setting:

$$\Delta = \frac{1}{J_{ms}} + \frac{1}{c_T V_e D} \qquad\qquad \left(\text{s.m}^{-2} \right)$$

$$\Sigma = \frac{V_s}{D_{me} \kappa} + \frac{V_s}{D_{ms}} + \frac{V_s}{c_T V_e D} - \frac{1}{c_T \kappa D} \qquad \left(\text{m.s.kmol}^{-1} \right)$$

After integration between the inlet and the outlet, we obtain:

$$c_{sP} = \frac{\Delta}{\Sigma} - \left(\frac{\Delta}{\Sigma} - c_{sE} \right) \exp\left(e_m J_s \Sigma \right)$$

Hereinafter, we shall suppose that we take J_s and J_e and the pore radius r_p. The liquid flux density is:

$$J_V = J_s V_s + J_e V_e$$

If we suppose that Poiseuille's law is applicable (the result found is fictitious if $r_p < 5 \times 10^{-9}$ m):

$$\frac{\Delta P}{e_m} = RT\frac{J_e}{D_{me}} = \frac{8\mu J_v}{r_p^2}$$

Thus, we have the diffusivity of the water in the membrane:

$$D_{me} = \frac{J_e r_p^2 RT}{8\mu J_v}$$

EXAMPLE.–

$J_e = 0.5 \times 10^{-3}$ kmol.m^{-2}.s^{-1} $\mu = 10^{-3}$ Pa.s $V_s = 0.035$ m^3.kmol^{-1}

$J_s = 4 \times 10^{-6}$ kmol.m^{-2}.s^{-1} $r_p = 0.5 \times 10^{-9}$ m $V_e = 0.018$ m^3.kmol^{-1}

$R = 8314$ J.kmol^{-1}.K^{-1} $e_m = 0.5 \times 10^{-6}$ m $T = 298.15$ K

$J_v = 4 \times 10^{-6} \times 0.035 + 0.5 \times 10^{-3} \times 0.018 = 9.14 \times 10^{-6}$ m.s^{-1}

$$D_{me} = \frac{0.5.10^{-3} \times 0.25.10^{-18} \times 8314 \times 298.15}{8 \times 10^{-3} \times 9.14.10^{-6}} = 4.24.10^{-9} \text{ m}^2.\text{s}^{-1}$$

$$\Delta P_e = \frac{8 \times 10^{-3} \times 9.14.10^{-6} \times 0.5.10^{-6}}{0.25.10^{-18}} = 1.462.10^5 \text{ Pa}$$

Hereafter, we consider that the total concentration c_T is constant, which greatly simplifies the calculations.

Let us now look for the diffusivity of the solute in the membrane for a concentration at the outlet equal to:

$c_s = 0.43 = J_s/J_v$

The additional data are as follows:

$c_{sE} = 1.3$ kmol.m^{-3} $D = 10^{-9}$ m^2.s^{-1} $V_e = 0.018$ m^3.kmol^{-1}

$$c_{sP} = 0.43 \text{ kmol.m}^{-3} \qquad\qquad V_s = 0.035 \text{ m}^3.\text{kmol}^{-1}$$

$$K = \frac{J_s}{J_e} = \frac{4 \times 10^{-6}}{0.5 \times 10^{-3}} = 8 \times 10^{-3}$$

$$c_T \# c_e \# \frac{J_e}{J_V} = \frac{0.5.10^{-3}}{9.14.10^{-6}} = 54.7 \text{ kmol.m}^{-3}$$

We suppose, after three tests and linear interpolations, that:

$$D_{ms} = 0.2275.10^{-11} \text{ m}^2.\text{s}^{-1}$$

$$\Sigma = \frac{0.035}{4.24.10^{-9} \times 8.10^{-3}} + \frac{0.035}{D_{ms}} + \frac{0.035}{10^{-9}} - \frac{1}{54.7 \times 8.10^{-3} \times 10^{-9}}$$

$$\Sigma = -1.219.10^9 + \frac{0.035}{0.2275.10^{-11}} = 1.416.10^{10} \text{ m.s.kmol}^{-1}$$

$$\Delta = \frac{1}{D_{ms}} + \frac{1}{54.7 \times 0.018 \times 10^{-9}} = \frac{1}{D_{ms}} + 10^9$$

$$\Delta = \frac{1}{0.2275.10^{-11}} + 10^9 = 4.4056.10^{11} \text{ s.m}^{-2}$$

$$\frac{\Delta}{\Sigma} = \frac{44.056}{1.416} = 31.1 \text{ kmol.m}^{-3}$$

We can verify that:

$$0.43 \# 31.1 - (31.1 - 1.3) \exp(0.5.10^{-6} \times 4.10^{-6} \times 1.416.10^{10})$$

We can now obtain the pressure drop across the membrane.

$$\Delta P = e_m RT \left(\frac{J_e}{D_{me}} + \frac{J_s}{D_{ms}} \right)$$

For water:

$$\Delta P_e = \frac{0.5.10^{-6} \times 24.35.10^5 \times 0.5.10^{-3}}{4.24.10^{-9}} = 1.42.10^5 \, Pa$$

For the solute:

$$\Delta P_s = \frac{0.5.10^{-6} \times 24.35.10^5 \times 4.10^{-6}}{0.2275.10^5} = 21.42.10^5 \, Pa$$

We now calculate the difference in osmotic pressures of the water between the two internal surfaces of the membrane. For this purpose, we shall calculate the molar fractions of water on those two surfaces.

$$x_{eE} = 1 - \frac{c_{sE}}{c_T} = 1 - \frac{1.3}{54.7} = 0.9762$$

$$x_{eP} = 1 - \frac{c_{sP}}{c_T} = 1 - \frac{0.43}{54.7} = 0.9921$$

The difference between the osmotic pressures is:

$$\Delta \Pi = \Pi_E - \Pi_S = \frac{8314 \times 293.15}{0.018} Ln\left(\frac{0.9921}{0.9762}\right)$$

$$\Delta \Pi = 21.87.10^5 \, Pa$$

This pressure corresponds fairly closely to the pressure due to the friction of the solute.

$$21.42 \# 21.87$$

The intrinsic transmission was:

$$T_{ri} = \frac{c_{sE}}{c_{sP}} = \frac{0.43}{1.3} = 0.33 = 33\%$$

A different calculation performed with different data and a 2% transmission confirmed the equality of $\Delta\Pi$ and ΔP_s.

2.2.9. *Practical equations*

Duroudier [DUR 99] gave the expression of the coefficients coming into play in these equations, on the basis of the Maxwell–Stefan equations.

$$J_s = (1-\sigma)J_v c_s - P_\mu \frac{d\mu_s}{dz} \tag{2.7}$$

$$J_v = -L_P \frac{d}{dz}(P - \sigma\Pi) \tag{2.8}$$

$$\sigma = \frac{c_T c_e V_e (V_e D_{em} - V_s D_{sm})}{\dfrac{D_{em} D_{sm}}{D} + c_T \left(c_e V_e^2 D_{em} + c_s V_s^2 D_{sm} \right)}$$

The coefficient σ is the "reflection coefficient"

$$P_\mu = \frac{V_e c_s c_T D_{sm} D_{em}}{\dfrac{D_{em} D_{sm}}{D} + c_T \left(c_e V_e^2 D_{em} + c_s V_s^2 D_{sm} \right)}$$

$$L_P = \frac{\dfrac{D_{sm} D_{em}}{D} + c_T \left(c_e V_e^2 D_{em} + c_s V_s^2 D_{sm} \right)}{RT \left(c_T + \dfrac{c_e D_{sm}}{D} + \dfrac{c_s D_{em}}{D} \right)}$$

However, for fairly dilute solutions and a notable rejection of the solute:

$$D_{sm} \ll D_{em}; \quad c_s \ll c_e; \quad D_{sm} \ll D$$

$$c_e V_e \#1 \quad c_T \# c_e$$

We then obtain:

$$\sigma = \frac{V_e}{V_e + \dfrac{D_{sm}}{c_T D}} \qquad P_\mu d\mu_s = \frac{RTV_e \Gamma_s dc_s}{\dfrac{1}{c_T D} + \dfrac{V_e}{D_{sm}}} \quad \text{or indeed} \quad \omega = \frac{RTV_e \Gamma_s}{\dfrac{1}{c_T D} + \dfrac{V_e}{D_{sm}}}$$

(μ_s is the chemical potential of the solute)

Equation [2.7] is then written more conventionally:

$$J_s = -\omega \frac{dc_s}{dz} + (1-\sigma) J_v c_s$$

The coefficient ω is the permeability of the membrane to the solute. As regards the expression of J_V, i.e. the practical equation [2.8]:

$$L_p \# \frac{V_e D_{em}}{RT} \quad \text{and} \quad J_V = \frac{V_e D_{em}}{RTe_m}(\Delta P - \sigma \Delta \Pi)$$

e_m: thickness of the membrane

2.2.10. *Expression of the rejection using the practical equations*

Remember the practical equation [2.7]:

$$J_s = -\omega \frac{dc}{dz} + (1-\sigma) J_v c$$

or indeed:

$$\left(c - \frac{J_s}{(1-\sigma) J_V} \right) dz = \frac{\omega}{(1-\sigma) J_V} dc$$

In view of the fact that $J_s / J_V = c_s$, let us integrate from the inlet (c_E) to the outlet (c_P) [SPI 66, KED 66].

$$\frac{e_m (1-\sigma) J_V}{\omega} = \ln\left(\frac{c_P (1-\sigma) - c_s}{c_E (1-\sigma) - c_s} \right) = \ln\left(\frac{\sigma c_P}{c_P - (1-\sigma) c_E} \right)$$

Let us introduce the intrinsic rejection R:

$$\frac{c_P}{c_E} = Tr = 1 - R$$

and set:

$$F = \exp\left(-\frac{J_v\left(1-\sigma\right)e_m}{\omega}\right)$$

We obtain:

$$F = \frac{\sigma - R}{\sigma(1-R)} \quad \text{i.e.} \quad R = \frac{(1-F)\sigma}{1-\sigma F}$$

We see that if $J_V \to \infty$, $R \to \sigma$.

σ *is the reflection coefficient* of the solute–membrane system.

Let us now calculate dR/dJ_V:

$$\frac{dR}{dJ_V} = \frac{dR}{dF} \times \frac{dF}{dJ_V} = \frac{\sigma(\sigma-1)}{(1-\sigma F)^2} \times \frac{(\sigma-1)e_m F}{\omega} = \frac{\sigma(1-\sigma)^2 e_m F}{(1-\sigma F)^2 \omega}$$

If we set:

$$P = \omega/e_m$$

When $J_V \to 0$, then $F \to 1$ and $\dfrac{dR}{dJ_V} \to \dfrac{\sigma}{P}$

and, if $J_V \to \infty$, then $F \to 0$ and $\dfrac{dR}{dJ_V} \to 0$

These results, concerning the intrinsic rejection, have been amply verified by experience.

2.2.11. *Pervaporation*

In this operation, the permeation of the liquid mixture is followed by the vaporization at the outlet from the membrane.

Kataoka *et al.* [KAT 91a, KAT 91b] use an overall diffusivity D in their expression of the diffusion. This diffusivity is constant over the thickness of the membrane. As a function of the composition of the liquid mixture at the inlet, they obtain a correct shape for:

– the overall mass flowrate across the membrane;

– the weight fractions of the components in the outgoing filtrate.

These authors offer a generalization, accepting that the activity coefficients are equal to 1 and they examine the following four situations:

– liquid or gaseous feed;

– liquid or gaseous filtrate.

This enables them to deal, at least in principle, with reverse osmosis and pervaporation. However, the results found by Kataoka *et al.* (1991a and 1991b) are merely qualitative.

Mulder and Smolders [MUL 84] gave a detailed examination of the expression of the activity of the compounds in a liquid mixture in a polymer membrane. In 1985, Mulder *et al.* brought into play the derivative of each activity with respect to the volumetric fractions of *all the components*. In addition, they used a pre-constructed membrane of successive, parallel layers, and thereby measured the concentration profile of each species inside the membrane. Their Figures 2.3 and 2.4 confirm their Figure 2.1. Flory [FLO 44] looked at the thermodynamics of polymers on contact with a liquid.

2.2.12. *Industrial applications of filtration on an electrically inert membrane*

By far the most important application is desalination of seawater (30 g of salt per liter) and brackish water at 2 g.L^{-1}, to make it into drinkable water at less than 0.5 g per liter.

Desalination takes place by reverse osmosis. Sirkar and Rao [SIR 81] proposed a method for calculations of such an installation. This method has two variants, so three distinct procedures. The fluxes of solvent and solutes are expressed by a solution–diffusion-type model. Applegate [APP 84] put forward procedural diagrams for reverse osmosis. The flowrates for reverse osmosis are around 15 L.day^{-1}.m^{-2}.bar^{-1}.

Aulas *et al.* [AUL 79] studied the recovery of metal cations by ultrafiltration of a solution of these ions. Marze [MAR 78] reviewed the applications of ultrafiltration.

Industrial membranes are *asymmetrical* in the sense that a thin film of less than 1μm in thickness performs the filtration but, to ensure the mechanical resistance of that film, it needs to be held, laying it on a film which is around 1mm thick.

Opong and Zydney [OPO 91] studied the transport of proteins across an asymmetrical membrane.

Soles *et al.* [SOL 82] looked at the crossing of a polyethylene membrane by gases.

2.3. The membrane and the solutes are electrically charged

2.3.1. *Prior conventions: notations*

σX represents the electrical charge of the membrane per unit volume, and σ will be equal to +1 or −1 depending on the sign of the charge.

We use the term "c-ions" to speak of co-ions – i.e. ions whose charge is of the same sign as that of the ions fixed to the membrane.

We also speak of "g-ions", which are counter-ions – i.e. ions whose charge is of opposing sign to that of the membrane. The prefix "g" comes from the German, *gegen*, which means "counter".

Remember that the intrinsic transmission of an ion is the ratio:

$$Tr_i = \frac{\text{concentration just before exiting the membrane}}{\text{concentration just after entering the membrane}}$$

Hereafter, we shall consider only those ions whose size is small in relation to the pores of the membrane. The coefficient λ in section 2.2.5 is equal to 0.

In addition, we consider only the concentrations expressed in relation to the liquid phase inside of the membrane.

In case we wish to bring in the concentrations outside of the membrane, we need to use an overall sharing coefficient which is defined as follows:

$$K_{GL} = \frac{c_{mi}}{c_{ei}} = \varepsilon \, k_{ED} \, k_{DO}$$

ε: neutral sharing coefficient (see section 2.2.5)

k_{ED}: sharing coefficient for dielectric exclusion (see section 2.3.2)

k_{DO}: Donnan sharing coefficient (see section 45.4)

The rejection is:

$$R_i = 1 - Tr_i$$

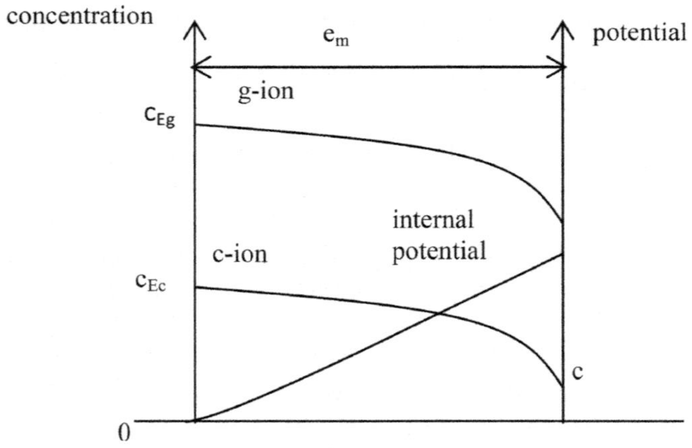

Figure 2.3. *Donnan effect on the inlet face.*
Negatively charged membrane

We have taken the precaution of not using the valence z_i in the expression of the dielectric exclusion. Thus, the value of the overall coefficient K_{GL} does not depend on the order in which the coefficients are multiplied.

c_{mi}: concentration in the membrane near to its surface

c_{ei}: concentration outside of the membrane near to its surface

2.3.2. Dielectrical exclusion

The relative permittivity of the polymers used to create membranes is about 3 to 5, so much less than that of water, which is around 80.

However, each ion induces on the surface of the medium of lesser permittivity – i.e. on the wall of the pores – a charge of the same sign as the ion considered which, therefore, is pushed back. This contributes to the exclusion of the ions outside of the pores.

Now let us set:

$$\alpha_i = \frac{L_B}{d_{pore}} = \frac{z_i^2 e^2}{4\pi\varepsilon_r\varepsilon_0 kTd_{pore}} = \frac{z_i^2 F^2}{4\pi\varepsilon_r\varepsilon_0 RTN_A d_{pore}} \qquad \text{(S.I.)}$$

Let us also set:

$$\lambda_0 = r_{pore}\kappa_0 \; ; \; \lambda_{pore} = r_{pore}\kappa_{pore}$$

L_B: Bjerrum length (Cabane and Henon, 2003) (m)

κ_0^{-1} : Debye length in the external solution on contact with the membrane (m)

κ_{pore}: Debye length in the liquid of the pores (m)

As we know from section 3.3.2, the Debye length is:

$$\kappa^2 = \frac{e^2}{\varepsilon_r\varepsilon_0 kT}\sum_j c_j z_j^2 \qquad \text{(S.I.)}$$

F: Faraday's constant ($F = 96.487 \times 10^6$.C.keq^{-1})

R: ideal gas constant (8314 J.kmol^{-1}.K^{-1})

T: absolute temperature (K)

k: Boltzmann's constant (1.38048×10^{-23} J.K^{-1}.molecule^{-1})

N_A: Avogadro's number (6.023×10^{26} molecule.kmol^{-1})

ε_0: permittivity of vacuum (8.854×10^{-12} C^2.m^{-1}.J^{-1})

The Helmholtz energy corresponding to the transport of an ion from the outside free solution into a pore is therefore:

$$\frac{W_i}{kT} = \frac{2\alpha_i}{\Pi} \int_0^\infty \left[\frac{K_0(k)K_1(v) - \overline{\beta}K_0(v)K_1(k)}{I_1(v)K_0(k) + \overline{\beta}I_0(v)K_1(k)} \right] dk + \alpha_i \left(\lambda_0 - \lambda_{pore} \right)$$

The variable v is an auxiliary variable defined by:

$$v = \sqrt{k^2 + \lambda_{pore}^2}$$

The ratio $\overline{\beta}$ is:

$$\overline{\beta} = \frac{k}{v}\beta = \frac{k}{v}\frac{\varepsilon_m}{\varepsilon_{solvent}}$$

I_0 and I_1: Bessel functions of order 0 and 1

K_0 and K_1: modified Bessel functions of order 0 and 1

ε_m and ε_{sol}: relative permittivities of the membrane and of the solvent

A description of the Bessel functions is to be found in Goude [GOU 65].

This expression of W is given by Yaroshchuk [YAR 00]. The same author also gives the equivalent result for a pore in the form of a crack.

The sharing coefficient is then written:

$$\phi = \frac{c_{me}}{c_i} = \frac{c_{ms}}{c_p} = \varepsilon(1-\lambda)^2 \exp\left(-\frac{W}{kT}\right)$$

The expressions involved in W_i show that the dielectrical exclusion increases greatly when the pore radius decreases.

It is important to note that if the solution contains ions of different valences, *their relative proportions will be different on either side of the surface of the membrane.*

2.3.3. *The fixed charge of the membranes. Influence of pH*

The fixed charge X is evaluated in kiloFaraday per m^3 of the liquid of the pores (of the impregnation liquid).

For the length L of a pore, there are two ways of expressing the fixed charge X. Let σ represent the charge in kiloFaraday per m^2 of pore surface.

$$L\pi r_p^2 X = 2\pi r_p L\sigma \quad \text{so} \quad X = \frac{2\sigma}{r_p}$$

r_p: mean pore radius (m)

where:

1kFaraday $= 1000$ Faraday $= 96.487\times10^6 C = 1kF$

IEP is the isoelectric point – i.e. the value pH_0 – for which the potential zeta is equal to zero (see Chapter 5 in this volume). We can see that the charge of the surface is negative if $pH > pH_0$. In this case:

– carboxylic acids shed protons, leaving a negative radical;

– anions such as OH^- (or others) are specifically adsorbed. These anions stem from the presence of salts in the solution.

If pH < pH$_0$, the surface charge is positive. The specific adsorption of cations can contribute, but they are more hydrated than the anions, and find it more difficult to approach the surface.

If pH = pH$_0$, then the fixed charge is often null and the rejection of the electrolytes by the membrane is also null.

We can find the value of the surface charge by knowing the distribution of the potential φ in a pore. Indeed, we know that Poisson's equation is written:

$$\Delta\varphi = \frac{1}{r}\frac{\partial}{\partial r}\left(\frac{\partial\varphi}{\partial r}\right) = -\frac{\rho}{\varepsilon\varepsilon_0} \quad \text{and by integrating from zero to } r_p \left(\frac{d\varphi}{dr_p}\right)_{r_p=0} = \frac{\sigma}{\varepsilon\varepsilon_0}$$

The value of the potential φ on the external surface of the Stern–Hamburg fixed layer is simply the potential zeta. We can see that there is no direct relation between the sign of zeta and the sign of σ. This relation of signs exists only between the sign of $\dfrac{d\varphi}{dr}$ and that of σ. However, the φ potential decreases in absolute value uniformly toward zero when the distance to the Stern layer increases indefinitely.

If we accept that σ is the total charge – i.e. the sum of the natural charge and the adsorbed charge – we can say that the signs of σ and ζ are identical. However, as the ion adsorption depends on the presence of salts in solution, σ and ζ do not go to zero for a fixed value pH$_0$ of the pH. In other words, the zero charge point is different to the isoelectric point.

NOTE.–

The hypothesis of the volumetric charge X is known as the TMS hypothesis (standing for Teorell–Meyer–Sievers; see Chapter 5 in Vol. 7). The hypothesis of the surface charge σ is incorrectly known as the "spatial charge" to some people.

Let us stress the fact that *it is the surface charge which is characteristic of the interaction (as we shall soon see) between the membrane and the solution impregnating it.* The volumetric charge, for its part, varies with the radius of the pore because [TAG 96, NAK 96]:

$$X = \frac{2\sigma}{r_p}$$

2.3.4. *Effectiveness coefficient of the fixed charge*

The nominal fixed charge is that which is deduced from the dissociation of the sulfonic- or carboxylic groups of the polymer constituting the matrix of the membrane. This nominal charge can be deduced by an acid/base chemical dosage according to the pH of the solution in contact with the membrane.

When an ionic solution passes through the membrane, the g-ion (which, for example, is a cation) is fixed to a portion of the negative charges, and neutralizes them so that the effective fixed charge on the wall of the pores becomes $\phi\sigma$ instead of σ.

This adsorption obeys Langmuir's law and we have:

$$\sigma = \phi\sigma_0 = \sigma_0 - \sigma_a \left(\frac{k_0 c_g}{1 + k_1 c_g} \right)$$

ϕ: effectiveness of the fixed charge

k_0 and k_1 have the dimensions of the inverse of a concentration $(m^3.kmol^{-1})$.

c_g is the concentration of the g-ions in the solution on contact with the pores.

The coefficients $(\sigma_a k_0)$ and k_1 depend on the nature of the cation in question. Therefore, it is helpful to recap the values of the ionic radii of the monovalent cations.

cation	Li	Na	K	Rb	Cs
$r(10^{-10}m)$	0.68	0.98	1.33	1.49	1.65

The smaller cations are hydrated most easily, which hinders them from approaching the wall of the pores, and thus from being adsorbed. The result of this is that the effectiveness ϕ of the fixed charge decreases in the following order [KIM 84].

Li > Na > K > Cs

The parietal charge σ (or volumetric charge X) is determined correctly by Teorell's method on the basis of a dialysis measurement and of the corresponding membrane potential. Originally, Meyer and Sievers had worked in the same way, with a graphical interpretation of their results. There are other methods in existence, but we shall not speak of them here.

The effectiveness of the fixed load may, according to some, drop as far as 0.2. It falls if we increase the ratio c_g/X because, in this case, adsorption increases [TSU 90].

NOTE.–

The fixed charge may be entirely due to the specific adsorption (non-ionic) of an ion by an initially neutral membrane [TAK 96, NAK 96].

NOTE.–

The fixed charge screens the interactions of the ions with the polarization charges, which decreases the dielectrical exclusion, and if the fixed charge increases greatly, the dielectrical exclusion may disappear completely.

When the ionic strength increases, the screening length $1/\kappa$ decreases, but this is accompanied by an increase in the apparent dielectric constant of the solvent, which slows the fall of $1/\kappa$.

2.3.5. *Mobility of ions in the membrane*

In principle, the ionic mobility is inversely proportional to the radius of the ion in question. However, for monovalent cations, the mobilities are classed as follows [KIM 84].

Li < Na < K < Cs

This shows that it is not the ionic radius which determines the mobility, but rather the degree of hydration of the ion, which increases if the ion itself shrinks. This conclusion is validated for hydrophilic membranes but, for a hydrophobic membrane composed of nylon 6, the classification of the mobilities is reversed, and decreases when the ionic radius increases [KIM 84].

When the pore radius decreases, it is the most heavily hydrated ions whose mobility decreases most greatly.

2.3.6. *The supplemented Nernst–Plank equation (SNP)*

Unlike the Maxwell–Stefan equation, this equation does not take account of any coupling between different solutes. The SNP equation, then, is a relation between the solute i and the solvent (generally water). Let us recap the Maxwell–Stefan equation (see section 2.2.8):

$$-c_i \frac{d\mu_i}{dz} - Fz_i c_i \frac{d\psi}{dz} = \left(x_e J_i - x_i J_e \right) \frac{RT}{D_i}$$

Suppose that:

$$J_V = \frac{J_e}{c_e}; \quad x_e \#1; \quad c_e \# c_T; \quad d\mu_i = RT \, d\ln c_i; \quad \frac{D_i z_i F}{RT} = u_i$$

After multiplication by D/RT, we obtain the SNP equation, which *would be better called the DMC equation* (for Diffusion, Migration, Convection)

$$J_i = -D_i \frac{dc_i}{dz} \qquad\qquad -u_i c_i \frac{d\psi}{dz} \qquad\qquad +c_i J_V$$

Diffusion   Electrical migration   Convection

Nernst [NER 88, NER 89] and Plank [PLA 90, PLA 90] introduced the first two terms on the right-hand side of this equation but, in order to take account of the operations of membrane filtration, it was necessary to introduce a convection term – i.e. the entrainment of the solute by the liquid current.

u_i: electrical mobility of the ion i $(m^2.s^{-1}.V^{-1})$

J_V: liquid flux density $(m^3.s^{-1}.m^{-2} - i.e.\ m.s^{-1})$

c_i: concentration of the solute i $(kmol.m^{-3})$

D: diffusivity of the solute i in the solvent $(m^2.s^{-1})$

The DMC equation has yielded excellent results. Note that it corresponds to a decoupling with the relation giving J_V as a function of dP/dz. In the cases of microfiltration and ultrafiltration, we can use Poiseuille's law for the liquid flow in a pore whose radius is r_{pore}, and write (see section 2.2.6):

$$J_V = \frac{r_{pore}^2}{8\eta} \frac{dP}{dz}$$

η: viscosity of the liquid phase (Pa.s)

The friction of the ions in the membrane is overlooked (although the same is no longer true for reverse osmosis).

$$RT / D_{mi} = 0$$

Besides the decoupling of the liquid flowrate, experience of the calculations shows the absence of coupling between two particles of solute. In particular, between two ions, the coupling is indirect, meaning that they are subject only to the electrical field.

Thus, whatever the natures of the ions i and j, we accept that:

$$RT / D_{ij} = 0$$

However, *this absence of interactions and mutual friction is no longer valid if the concentrations c_i increase significantly.*

The electrical potential gradient is obtained by writing that:

$$0 = \Sigma z_i J_i = -\Sigma z_i D_i \frac{dc_i}{dz} - \sum_i z_i u_i c_i \frac{d\psi}{dz} + \sum_i z_i c_i J_V$$

The neutrality of the solution impregnating the membrane is written:

$$\sum_i z_i c_i + \sigma X = 0 \qquad\qquad \text{(here, } \sigma \text{ is equal to +1 or to -1)}$$

Hence, finally:

$$\frac{d\psi}{dz} = \frac{1}{\sum_i u_i z_i c_i}\left(-\sigma X\, J_V - \sum_i z_i D_i \frac{dc_i}{dz}\right)$$

However, we always have:

$$z_i u_i > 0 \qquad \sum_i z_i c_i = -\sigma X = \text{const.} \quad \text{so} \quad \sum_i z_i \frac{dc_i}{dz} = 0$$

As the D_i values are all of the same order of magnitude, the second term in the parenthesis is smaller than the first. Consequently, the sign of $d\psi/dz$ is opposite to that of X.

2.3.7. *Numerical resolution of the DMC system of equations*

We know that the DMC equation is:

$$J_i = -D_i \frac{dc_i}{dz} - u_i c_i \frac{d\psi}{dz} + c_i J_V \qquad\qquad [2.9]$$

and that the gradient of electrical potential is:

$$\frac{d\psi}{dz} = \frac{-1}{\sum_i u_i z_i c_i}\left[\sigma X J_V + \sum_i z_i D_i \frac{dc_i}{dz}\right] \qquad\qquad [2.10]$$

If we replace $d\psi/dx$ in equation [2.9] with its value given by equation [2.10], we obtain a system of n linear equations in terms of dc_i/dz (the value of i ranging from 1 to n), where n is the number of solutes. The equations are of the form:

$$\frac{dc_i}{dz} = F\left(c_1, \ldots, c_i, \ldots, c_n\right)$$

Such a system of equations can easily be solved using the 4th-order Runge–Kutta method as described in Appendix 3, because we know the concentrations on entering the membrane.

Thus, we can, in principle, obtain the internal concentration profiles. However, the flux densities J_i play a part in all the equations, and these quantities are unknown, *a priori*.

If c_{Pi} is the concentration of the permeate *just after exiting* the membrane, we can write:

$$J_i = J_V c_{Pi}$$

The c_{Pi} values are obtained by applying the Donnan equilibrium on both sides of the face of exit from the membrane (see section 3.4.2). An initial calculation of the concentration profiles can be obtained by making all J_i values equal to zero, except for that of water. The next rounds of calculation will be performed with the values obtained for the c_{Pi}. Two or three iterations are sufficient.

NOTE.–

The Donnan equilibrium between the wall of the membrane and the external solution is described by Gavach [GAV 77], by Bergsma and Kruissink [KRU 61] and is used by Meyer and Sievers [MEY 36, SIE 37].

NOTE.–

The 4th-order Runge–Kutta method can also be used if we increase the complication of the DMC equation with the binary interaction terms which play a role in the Maxwell–Stefan equations. Let us give some indications here:

$$c_e V_e + \sum_i c_i V_i = 1 \quad (V_i \text{ is the molar volume of the solute i and } V_e \text{ that of water})$$

$$c_i = c_T x_i \quad \text{and} \quad c_T = \frac{1}{\sum x_i V_i + x_e V_e} \quad \text{and} \quad x_e + \sum_i x_i = 1$$

In addition, if the behavior of the solutes is not ideal:

$$d\mu_i = RT\left(1 + \frac{d\ln\gamma_i}{d\ln x_i}\right)d\ln x_i = RT\Gamma_i d\ln x_i$$

Finally, the system of equations to be integrated is of the form:

$$\frac{dx_i}{dz} = F\left(x_1, \ldots, x_i, \ldots, x_m\right)$$

NOTE.–

As regards non-electrolytes, the chemical potentials can be expressed by Wilson's method (see Chapter 3). However, if we wish to associate the polymer in the membrane with the calculations of chemical potentials, we can use Flory's [FLO 44] method and Huggins' [HUG 41]. An example of the use of this method is given by Mulder *et al.* [MUL 84].

2.3.8. *The approximate calculations presented by Dresner [DRE 72]*

In the DMC equation, Dresner considers that the molar flux density J_i is very slight in comparison to the other terms, and he deduces the profile (over the thickness of the membrane) of the internal potential.

When we know that profile, the complete DMC equation gives us the concentration profile for the c-ions.

In addition, Dresner [DRE 72] gives the necessary and sufficient condition for a single g-ion to pass through the membrane, as certain researchers have noted.

Finally, Dresner gives us the equation for the calculation of the J_i of the c-ions.

NOTE.–

The justification for Dresner's [DRE 72] integration is found in Spiegel [SPI 74].

2.3.9. *Influence of pore radius*

The layer of g-ions is in direct contact with the wall of the pore, simply through electrostatic attraction. The thickness of the layer of g-ions is near to the Debye length $1/\kappa$.

The c-ions occupy the central part of the pore, and that portion decreases if the pore diameter decreases. However, the progression of the electrolyte through the membrane is determined by the cross-section available to the c-ions. Hence, the rejection increases if the pore diameter decreases. Let us examine this in greater detail.

The diffuse layer attached to the wall of the pores is thicker when the ionic strength is low (see section 3.3.2). This thickness is $1/\kappa$. In order for diametrically-opposed double layers to overlap, we must have:

$$r_{pore} < 2,2 / \kappa$$

However, if the double layers overlap, the rejection of the axial c-ion is increased, and so too is the rejection of the electrolyte.

In other words, rejection increases if the diameter of the pores decreases. In addition, dielectric exclusion increases greatly with the fineness of the pores (see section 2.3.2), which also contributes to the increase of the rejection.

If the ion concentration falls, the swelling of the membrane increases and, if the swelling increases, the permeability available for the liquid phase also increases.

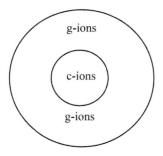

Figure 2.4. *The ions in a pore*

2.3.10. *Transport of electrolytes across a charged membrane*

To clarify our ideas, *we shall suppose that the membrane is negatively charged* by affixed groups. A negative charge is by far the most common situation, and it is due to the presence of carboxylic- and/or sulfonic radicals.

In the discussion below, we can assume that the electrical potential rises from zero to ΔV between entering and exiting the membrane – i.e. in the direction of the liquid flux. This increase in electrical potential creates an electrical field E opposite to the movement of the liquid. Three consequences arise from this:

1) the g-ions (positive) are entrained by the liquid flux, but this motion is hindered (or "compensated", to use the dedicated term) by the electrical field E. Finally, the g-ions, which are present in a high concentration, progress slowly with the liquid, owing to that compensation;

2) the c-ions (negatives) are entrained both by the flux J_V and by the electrical field E. Thus, these two forces are added together, and the c-ions move quickly, but it must not be forgotten that their concentration is low;

3) it stems from these inequalities of velocity and concentration that the neutrality of the electrical flux is preserved. Let v_g and v_c be the velocities in relation to the membrane of the g-ions and c-ions. Then, the neutrality is written:

$$z_g v_g c_g + z_c v_c c_c = 0$$

Here, $z_g > 0$ and $z_c < 0$.

2.3.11. *Structure of pores and rejection*

Hoffer and Kedem [HOF 67], and in particular Wang *et al.* [WAN 95], arrived by calculation at the following conclusions:

1) the concentration c_c of the c-ion decreases linearly across the thickness of the membrane if the flux density J_V is moderate, but if J_V has a high value, the concentration c_c decreases exponentially. Remember that, in the membrane, the concentration c_g of the g-ion is:

$$z_g c_g = z_c c_c + X$$

2) the radial profile of the flowrate of c-ions is maximal on the axis of a pore and tends toward zero on its wall. The profile of flow of the g-ions, which is significant near to the axis, becomes zero before reaching the wall and then becomes increasingly negative as it approaches the wall. The result of all of this is that the *flow* passing through the membrane is electrically neutral;

3) the g-ions are accumulated in the vicinity of the wall and may be partially adsorbed, which causes a sometimes-unpredictable decrease of the fixed charge;

4) the velocity profile of the liquid phase in the pore is very close to a parabola, as predicted by Poiseuille's law, and this holds true for a pore diameter of the order of 2mm, which corresponds to half a dozen diameters of the water molecule. In particular, in nanofiltration, we can accept a Poiseuille profile;

5) the rejection of a membrane increases with the reflection coefficient σ and decreases with the permeability to the solute ω. When one of the (the c-ion OH⁻ or the g-ion H⁺) has high mobility, the rejection may be negative. Rejection generally increases with the ratio X/c, where X is the fixed charge and c is the concentration of the prefilter solution. The rejection decreases if the ratio r_p/λ_D increases, where $\lambda_D = 1/\kappa$ is the Debye length (see section 2.3.2). Remember that if the ionic concentration increases, the length λ_D decreases;

6) the rejection increases with J_V. If J_V is small, diffusion predominates over convection and, if J_V increases, the opposite will be true, and the rejection will tend toward σ (see section 2.2.10);

7) Tsuru *et al.* [TSU 91] showed, with regard to the valences of the ions, that the intrinsic rejection increases if:

– the valence of the c-ion increases,

– the valence of the g-ion increases,

– the mobility of the c-ion decreases,

– the mobility of the g-ion decreases.

Thus, SO_4^{--} and CrO_4^{--} are better rejected than NO_3^{-}. Indeed, the exclusion of bivalent ions by the fixed charge is stronger than that of monovalent ions.

Similarly, H^+ and OH^- ions, which are highly mobile, give rise to a negative rejection, until it reaches a minimum and then recovers once more when we increase the fixed charge.

The Ca^{++} and Mg^{++} ions are better rejected than the sodium ions Na^+.

8) Yaroshchuk [YAR 00] examined what happens if, with a dominant g-ion, we add an extra g-ion. If the extra g-ion is more mobile, the conductivity of the membrane increases, the internal electrical field decreases and we see a reduction in the rejection of the dominant g-ion. On the other hand, if the extra g-ion is less mobile, the rejection of the dominant g-ion increases. Thus, the rejection of K^+ improves if we add the less-mobile g-ion Li^+.

However, when we apply a g-ion of low mobility, sooner or later the ionic strength will increase and screen the action of the fixed load, so that the rejection of the dominant g-ion will decrease after passing through a maximum.

2.3.12. *At least three different g-ions*

When we are dealing with a single binary electrolyte, the flux of the g-ion must be stoichiometric in relation to the flux of c-ions. For this purpose, the convective flux of the g-ion is compensated by the electrical flux of that same g-ion.

When other g-ions of different mobilities appear in the feed, such compensation is impossible for all the g-ions and the electrical field acquires an intermediary value, so we have:

– undercompensation for the less-mobile g-ions. This is so-called "normal" behavior;

– overcompensation for the more-mobile g-ions. This is "abnormal" behavior. The rejection of these ions may even become negative.

For abnormal g-ions, overcompensation corresponds to a negative entrainment in the direction of the liquid stream. Their rejection can only tend toward their reflection coefficient if the liquid flux increases, because this would mean a negative concentration in the permeate.

Note that it is not the ions which are normal or abnormal but is rather the solution in its entirety. In addition, the c-ions are always normal because their convective and electrical fluxes are in the same direction.

Let us now look again at "normal" g-ions – that is, ones which exhibit a powerful negative rejection because they are much less entrained by the electrical field. These ions could leave c-ions to accompany them across the membrane and thus maintain the electric current at zero. However, this concern is unjustified if the molar fraction m of the less-mobile g-ions is lesser.

This fraction is defined by:

$$m = \frac{\text{concentration of less-mobile g-ions}}{\text{concentration of all g-ions}}$$

In summary:

Low m value: the behavior of the g-ions is normal

High m value: the behavior of certain g-ions is "abnormal" because *the electrical field increases with the proportion of less-mobile g-ions*

However, if the fixed charge increases, a lower value of m is sufficient to create overcompensation of the more-mobile ions and their abnormal behavior. Conversely, if the more-mobile g-ion has a double charge which partially neutralizes the fixed charge X, the abnormal behavior will manifest itself for higher values of m.

2.3.13. *Additional information*

Mackie and Meares [MAC 55] studied the diffusive flux of an electrolyte across a cation-exchange membrane.

Bergsma and Kruissink, in their publication, gave a thorough review of the state of knowledge (very advanced even at that point) up until 1961.

An exposition of the *MST theory (Meyer, Sivers, Teorell)* is found in Meyer and Sivers [MEY 36, SIV 37] and in Teorell [TEO 35]. This theory *describes the crossing of an ion membrane by g-ions*.

2.4. Electrohydraulic values

2.4.1. *Electro-osmotic parameters: specific cases*

The volumetric flux density J_V and the current density i passing through a membrane are coupled by electrokinetic phenomena.

$$J_V = \omega \frac{\Delta P}{e_m} + \alpha \frac{\Delta V}{e_m} \qquad [2.11]$$

$$i = \alpha \frac{\Delta P}{e_m} + \beta \kappa_\infty \frac{\Delta V}{e_m} \qquad [2.12]$$

e_m: thickness of the membrane (m)

ΔV: electrical potential difference on crossing the porous barrier (or the membrane) (V)

ΔP: pressure difference on crossing the membrane (Pa)

ω: hydraulic permeability coefficient

κ_∞: specific conductivity of the solution of the membrane ($\Omega^{-1}.m^{-1}$)

α: electro-osmotic coefficient ($A.s^2.kg^{-1}$ or $m^2.s^{-1}V^{-1}$)

(indeed $[AV] = kg.m^2.s^{-3}$)

β: coefficient of increase in conductivity (dimensionless)

The coefficient α ensures the coupling between hydraulics and electrokinetics. It is the electro-osmotic coefficient. In view of Onsager's reciprocity relations, it is the same coefficient α which is involved in both equations.

Let us now examine a few particular cases:

1) Flow stream. Suppose that ΔP imposes a value of J_V and that the two sides of the membrane are in short-circuit, meaning that $\Delta V = 0$. In view of

equation [2.12], a current arises. That current is the flow current, whose density is:

$$i_e = \alpha \frac{\Delta P}{e_m}$$

2) Flow potential. Again, suppose that ΔP imposes a value of J_V, but that the electric circuit linking the two sides of the membrane is open. Then, we see the appearance of a potential difference between the two faces of the membrane. This is the flow potential ΔV_e for $i = 0$.

$$\Delta V_e = -\frac{\alpha}{\beta \kappa_\infty} \Delta P$$

3) Increasing the viscosity. Let us calculate J_V for $i = 0$. The potential ΔV_e then plays a role in equation [2.9], which is written:

$$J_{V_1 i=0} = \frac{\omega \Delta P}{e_m}\left(1 - \frac{\alpha^2}{\omega \beta \kappa_\infty}\right)$$

Naturally, for $\Delta V = 0$, we obtain a simple hydrodynamic flow.

$$J_{V_1 \Delta V=0} = \frac{\omega \Delta P}{e_m}$$

Thus, we have the ratio:

$$\rho = \frac{J_{V_1 \Delta V=0}}{J_{V_1 i=0}} = \left(1 - \frac{\alpha^2}{\omega \beta \kappa_\infty}\right)^{-1}$$

As the flux is inversely proportional to the viscosity:

$$\frac{\eta_{app}}{\eta_0} = \frac{\eta_{i=0}}{\eta_{\Delta V=0}} = \frac{J_{V_1 \Delta V=0}}{J_{V_1 i=0}} = \left(1 - \frac{\alpha^2}{\omega \beta \kappa_\infty}\right)^{-1} = \rho$$

The ratio ρ is the ratio of the apparent viscosity – i.e. the viscosity in the presence of electrical phenomena – to the viscosity in the absence of these phenomena. The apparent viscosity is always greater than the viscosity of the solution outside of the membrane.

2.4.2. *Mobility, diffusivity, conductivity, transport number*

By definition, the conductivity in the pores is expressed by:

$$g = F \sum_i z_i c_i u_i \quad \text{where } z_i u_i > 0 \quad \left(\Omega^{-1} m^{-1} = S\,m^{-1} \right)$$

F: Faraday's constant (96.487×10^6 C.kequiv.$^{-1}$)

z_i: valence of the ion (this valence may be positive or negative)

c_i: concentration of the ion (kmol.m^{-3})

u_i: mobility of the ion (m².s^{-1}.V^{-1} = m.s^{-1}(V.m^{-1})$^{-1}$)

From the discussion in section 4.6.5, we know that:

$$u_i = \frac{z_i D_i F}{RT}$$

Hence:

$$g = \frac{F^2}{RT} \sum_i z_i^2 D_i c_i$$

R: ideal gas constant (8314J.kmol^{-1}.K^{-1})

T: absolute temperature (K)

D_i: diffusivity of the ion i (m².s^{-1})

Note that the transport number of the ion i is inside the membrane:

$$t_i = \frac{z_i^2 c_i D_i}{\sum_j z_j^2 c_j D_j} = \frac{z_i c_i u_i}{\sum_j z_j c_j u_j}$$

If we bring into play the sharing coefficients $K_i = \dfrac{c_i}{c_{ei}}$, we of course obtain the following (the index e characterizes the exterior of the membrane):

$$t_i = \frac{z_i K_i c_{ei} u_i}{\sum\limits_j K_j z_j c_{ej} u_j}$$

c_{ei}: concentration in the outside solution at equilibrium with the pores

2.4.3. *Measuring the potential* ζ *(see section 5.1.2)*

Levine *et al.* [LEV 75] integrated the Poisson–Boltzmann equation and the Navier–Stokes equation. Thus, they are able to obtain:

$$\zeta = \varphi_{r=r_p}$$

These calculations were confirmed by Van Keulen and Smit [KEU 92], who evaluated the relevance of the various expressions of the φ potential obtained by different authors.

Christoforou *et al.* [CHR 85] used an approach similar to that of Levine *et al.*, to supplement Smoluchowski's expression of the flow potential. Unfortunately, these authors limit their examination to an electrolyte where both the ions have the same valence, in terms of absolute value.

From the experimental point of view, we cite:

– Elimelech *et al.* [ELI 94] studied the influence of the pH and the saline concentration on the potential ζ;

– Nyström *et al.* [NYS 94] describe the device for measuring the flow potential on crossing a membrane;

– Werner *et al.* [WER 95] describe the device used for sweeping the surface of a planar membrane for measuring the flow potential. The same is true of Peeters *et al.* [PEE 99];

– Hidalgo-Alvarez *et al.* [HID 85] measured the flow potential and deduced the zeta potential during the sedimentation of a dispersion and found the same value on the crossing of a porous washer. However, these measurements pertained to high values of the zeta potential.

2.4.4. *The flow potential [SMO 03]*

The flow potential E_e is given in SI units:

$$E_{eSI} = \frac{\varepsilon \varepsilon_0 \zeta}{\eta \kappa} \Delta P$$

ε_0: permittivity of a vacuum $((36 \pi \times 10^9)^{-1} \ C.V^{-1}.m^{-1})$

ζ: zeta potential (volts)

η: viscosity (Pa.s)

κ: conductivity $(C.V^{-1}.s^{-1}.m^{-1})$

ε: relative permittivity of the liquid

From this, we deduce (see Appendix 2), if we want to switch to CGS EM units:

$$E_{eSI} = \frac{\varepsilon \zeta \Delta P \times 10 \times 910^9}{36 \pi 10^9 \times \kappa \times 10 \times \mu} = \left(\frac{2\Delta P}{4\pi \mu \kappa} \right) \zeta_{CGSEM}$$

It is in this form that Smoluchowski [SMO 03] demonstrated the relation giving the flow potential E_e.

EXAMPLE.–

Calculation of a flow potential in the International System (SI):

$\zeta = 0.05V$ $\mu = 10^{-3}$ Pa.s $\kappa = 0.6$ S.m^{-1}

$\Delta P = 10^5$ Pa $\varepsilon_0 = \left(36\pi 10^9\right)^{-1}$ $\varepsilon_r = 80$

In the International System:

$$V_e = \frac{10^5 \times 80 \times 0.05}{36\pi 10^9 \times 10^{-3} \times 0.6} = 0.0058V$$

In the CGS EM system (see Appendix 2):

$$\zeta = \frac{0,05}{300} \text{ volt stat} \qquad \Delta P = 10^6 \text{ barye} \quad \varepsilon = 80$$

$$\eta = 10^{-2} \text{ poise} \qquad\qquad\qquad \kappa = 0,6 \times 9.10^9$$

Indeed:

$$[\kappa] = \frac{I}{Vm} \text{ and } \frac{\kappa_{CGSEM}}{\kappa_{SI}} = \frac{3 \times 10^9 \times 300}{100} = 9 \times 10^9$$

$$V_e = \frac{10^6 \times 80 \times 0.05 \times 300}{4\pi \times 0.01 \times 0.6 \times 9 \times 10^9 \times 300} = 0.0058V$$

2.5. Fouling of membranes

2.5.1. *Introduction*

Fouling reduces the flowrate of permeation across the membrane. In practical terms, we can never accept a drop in flowrate of more than 10% or, at worst, in certain very specific cases, 30%. However, some of that drop in flowrate may be due to a compression of the membrane, which decreases its porosity, resulting from an excessive permeation pressure.

There are three sorts of fouling agents:

– organic (macromolecules, biological waste). Proteins bind strongly to hydrophobic membranes (polyethylene, polypropylene, polytetrafluoroethylene) whereas hydrophilic membranes (derivatives of cellulose, polyamides) are less susceptible to adsorption;

– inorganic (scale formation). The salts precipitate, forming a hard film, resistant to entrainment by the liquid. This film is known as "tartar" (or "scale"), which has little to do with tartaric acid.

– particles and various types of waste which can obstruct the pores. However, IO- or pervaporation membranes do not have pores.

NOTE.–

At the entry point to the membrane, where the concentration is highest (because of the polarization), the aggregation of macromolecules or the precipitation of salts are the most likely. The aggregates and certain precipitates behave like small colloids, meaning that their size is no greater than 0.5μm.

2.5.2. Adsorption of organics and colloids

The size of colloids is between a few nm and a few microns. These may be organic (whether aggregated or otherwise) or inorganic (clay, silica, salts, oxides, sulfides). Half of the causes of fouling of membranes are organic in nature.

They may occur:

– by the blocking of the pores if their size is slightly less than the diameter of the pores. Blocking is encouraged by narrowing and restrictions of the pores;

– by adsorption on the walls of the pores if the particle size is much less than the pore diameter. If adsorption is fast, it is a process of equilibrium between the concentration of the liquid and the degree of coverage of the wall. The adsorption isotherm is generally Freundlichian.

If e_D is the thickness of the adsorbed layer, the porosity ε becomes:

$$\frac{\varepsilon}{\varepsilon_0} = \frac{\pi\left(r_{p0} - e_D\right)^2}{\pi r_{p0}^2} = \left(1 - \frac{e_D}{r_{p0}}\right)^2$$

e_D: thickness of the adsorbed layer (may be of the order of 1.5μm): m

When the pores become clogged, their number n does not vary.

The flowrate for the Poiseuille flow in a pore with radius r_p is:

$$Q_u = \frac{\pi \, r_p^4 \, \Delta P}{8 \mu \, Z}$$

The flowrate per m² of membrane is:

$$J_V = \frac{n \, \pi \, r_p^4 \, \Delta P}{8 \mu \, Z}$$

where n is the number of pores per m² of membrane. The initial flowrate Q_0 across the membrane is:

$$J_{V0} = \frac{n \, \pi \, r_{p0}^4 \, \Delta P}{8 \mu \, Z}$$

Thus:

$$\frac{J_V}{J_{V0}} = \left(\frac{r_p}{r_{p0}} \right)^4 = \left(1 - \frac{e_D}{r_{p0}} \right)^4$$

We can see that the volumetric flux density J_V decreases rapidly with adsorption.

Let us now examine the chemistry of fouling by organic substances and colloids.

Multivalent cations (Fe^{+++} and Al^{+++}) are ligands for organic particles which, for their part, are charged negatively, which facilitates fouling but creates porous deposits.

The presence of Ca^{++} ions causes the formation of ball-shaped macromolecules and has an effect similar to that of trivalent ions.

The electrical charge of the colloids stems from:

– the specific adsorption of ions (non-electrostatic adsorption);

– the isomorphic substitution of surface ions by ions from the solution (clays, minerals such as oxides or sulfides);

– an anisotropic crystalline structure (clay);

– the ionization of surface groups (amines, carboxylic- or sulfonic acids).

Colloids are deposited on the surface of the membranes (preferentially in the hollows).

The deposition results from the opposing actions of the permeation pressure and the repulsion of the electrical double layers. However, other forces come into play:

– forces of repulsion between hydrated surfaces;

– forces of attraction between hydrophobe surfaces;

– repulsive forces of steric hindrance for the penetration of the polymers.

Organics and colloids can form gels if the solubility of the non-crystalline particles is surpassed – i.e. at the surface where the particles enter the membrane (polarization layer).

2.5.3. *Partial blocking (braking) of the filtrate in the pores by adsorption*

Braking is due to the irreversible adsorption of solute on the walls of the pores. This adsorption can correspond to an isotherm:

$$\left.\begin{array}{l} - \text{Freundlichian } c_P^x = \alpha c_s^\beta, \text{ or else } c_s^* = \left(\dfrac{c_P}{\alpha}\right)^{1/\beta} = f(c_P) \\[4mm] - \text{Langmuirian } c_P^* = \dfrac{\alpha c_s}{1+\beta c_s} \text{ or else } c_s^* = \dfrac{c_P}{\alpha - \beta c_P} = f(c_P) \end{array}\right\} \qquad [2.13]$$

As the liquid progresses in a pore of the length dz, the surface concentration rises from zero to the value c_P. Simultaneously, the local concentration of the liquid over the length dz passes from the value c_{s0} to the value $c_s^*(c_P)$ because we suppose that the equilibrium is quickly established.

Per unit time, the material balance is written:

$$\pi\, r_p^2\, q\left(c_{so} - c_s^*\right) = 2\,\pi\, r_p\, c_p\, \frac{dz}{d\tau} = 2\,\pi\, r_p\, c_p\, q$$

$$\frac{r_p}{2}\left(c_{s0} - c_s^*\right) = c_p \qquad\qquad [2.14]$$

q: liquid flowrate expressed in relation to the cross-section of the pore $(m.s^{-1})$

c_s: concentration of solute in the liquid $(kmol.m^{-3}$ or $kiloparticles.m^{-3})$

c_p: surface concentration of the solute on the wall of the pores $(kmol.m^{-2}$ or $kiloparticle.m^{-2})$

In view of equation [2.13], equation [2.14] enables us to calculate c_p:

$$c_{s0} - f\left(c_p\right) = \frac{2}{r_p} c_p \quad \text{and therefore} \quad c_p = g\left(c_{s0}\right)$$

After that deposition of solute, the liquid thus impoverished of solute can no longer deposit anything, because then, if an additional deposit were to occur, we would have:

$$c_s < c_s^*$$

Finally, a wave of concentration c_p propagates in the pore at a velocity $dz/d\tau = q$. The total adsorption time corresponds to the time taken by the filtrate to cross the membrane. It is a very short period of time. Let $c_p^{(1)}$ represent the surface concentration obtained after the passage of that wave.

Now suppose that we continue to feed filtrate into the pore. Equation [2.14] then becomes:

$$\frac{r_p}{2}\left(c_{s0} - c_s^{*(2)}\right) = c_p^{(2)} - c_p^{(1)}$$

Hence, we have the value of $c_p^{(2)} > c_p^{(1)}$.

After 4 or 5 passages of the liquid, $c_s^{*(u)}$ has become very close to c_{s0}, if not equal to c_{s0}, and the concentration c_p is then:

$$c_p = c_p^* (c_{s0})$$

This adsorption results in a decrease in the pore radius and, consequently, a decrease in the filtration flowrate for a constant value of ΔP. We can write:

$$\Delta r_p = V_s c_p^*$$

V_s: molar volume or particle volume of the solute ($m^3.kmol^{-1}$ or indeed $m^3.kiloparticle^{-1}$)

NOTE.–

A slightly different interpretation of the dynamics of pore-clogging is to accept that a constant fraction X of the solute introduced is deposited on the walls:

$$L\, 2\, \pi\, r\, \frac{dr}{d\tau} = -Xc_s q \quad \text{Hence} \quad 2r\frac{dr}{d\tau} = -\frac{Xc_s q}{n\, L\, \pi}$$

However, according to Poiseuille's law:

$$\frac{Q}{n} = q = \frac{\pi\, r^4 \Delta P}{8\, \mu\, L} \quad \text{d'où} \quad \sqrt{\frac{Q}{Q_0}} = \frac{r^2}{r_0^2}$$

and, by differentiating, $2r\dfrac{dr}{d\tau} = \dfrac{r_0^2}{2\sqrt{q_0 q}} \times \dfrac{dq}{d\tau}$

Thus:

$$\frac{-Xc_s q}{L\pi} = \frac{1}{2}\frac{r_0^2}{\sqrt{q_0 q}}\frac{dq}{d\tau} \quad \text{or indeed} \quad -\frac{2Xc_s}{L\pi r_0^2}\sqrt{q_0} = \frac{1}{q^{3/2}}\frac{dq}{d\tau}$$

We set:

$$k = \frac{2Xc_s}{L\pi r_0^2}$$

Let us integrate:

$$-\sqrt{q_0}\, k\tau = -2q^{1/2} + 2q_0^{-1/2}$$

If we multiply by $\frac{1}{2}\sqrt{q_0}$, we obtain the law of evolution of the flowrate as a function of time τ:

$$\sqrt{\frac{q_0}{q}} = 1 + \frac{k\tau q_0}{2}$$

2.5.4. *Scale formation by inorganic substances*

Scale is a hard, adherent deposit. The main scaling agents are:

– ferric hydroxide and aluminum hydroxide;

– silica. The solubility of amorphous silica in water is 100–150 mg.L^{-1} for $5 < pH < 8$. It increases with temperature if the pH goes beyond 9.5;

– calcium- and magnesium salts. The main one is gypsum ($CaSo_42H_2O$), on whose solubility the pH and the temperature have little influence. The solubility of calcium phosphates decreases greatly with the temperature at acidic pH values. We must also cite $CaCO_3$ and $MgCO_3$.

Let us examine the mechanism of scale formation:

In Figure 2.5, we see a metastable zone which corresponds to oversolubility, but this zone is poorly defined, and is delimited by the dashed curve in the figure.

The apparition of crystalline germs (known as nucleation) take place when we reach a certain degree of oversaturation S. We distinguish between three sorts of nucleation:

– homogeneous, within a clean solution;

– heterogeneous, in a solution containing dust;

– superficial, on contact with a surface, particularly if that surface is rough. The rugosity of the membranes is situated between 10 and 50 nm, and on rare occasions, can reach 500 nm.

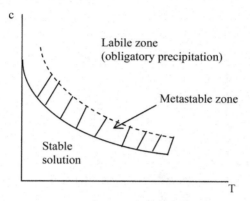

Figure 2.5. *Concentration/temperature diagram*

Practice shows that, for oversaturations:

$$S_{homo} > S_{hetero} > S_{surface}$$

Salts with low solubility are those which precipitate most easily.

Scale formation begins with the appearance, on the surface, of "islands" or "hills" which grow in height but especially in width – particularly if the rate of sweeping of the surface is low.

2.5.5. *Total blocking of the pores by the particles*

This phenomenon occurs all the more quickly when the diameter of the particles approaches and then surpasses the diameter of the pores.

If n is the number of open pores per unit surface of the membrane, we can write:

$$\frac{dn}{d\tau} = -J_v \frac{n}{k}$$

k: coefficient (m)

J_V: flux density in terms of volume ($m^3.m^{-2}.s^{-1}$)

Let us integrate:

$$\frac{n}{n_0} = \exp\left(-\frac{J_V \tau}{k}\right)$$

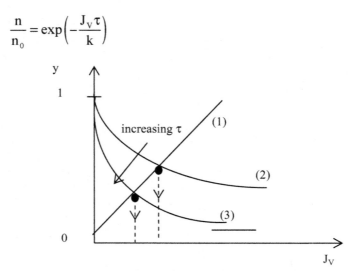

Figure 2.6. *Graphical determination of J_s*

If we take a particular value of ΔP, with the porosity ε being proportional to n, the flowrate J_V is also proportional to n, and the equation giving J_s as a function of the time becomes:

$$J_V = J_{V0} \exp\left(-\frac{J_V \tau}{k}\right)$$

This equation allows us to calculate J_V as a function of the time and as a function of the values of k and J_{V0}. We can use the graphical resolution in Figure 2.6.

J_V is the abscissa of the point of intersection of the straight line (1) y = J_V/J_{V0} and the curve y = exp ($-J_V\tau/k$).

From curve (2) to curve (3), the time has increased and J_V has decreased.

After a sufficiently long period of time, the blocking becomes total for all the pores and the membrane is impermeable to the solution.

2.5.6. *Partial blocking by the particles (hindrance by arching): formation of filtration cake*

The phenomenon exhibits like the total blockage, with the difference that the free section available for the passage of the liquid is the sum of those of:

– pores which are not yet partially obstructed:

$$n_1 = n_0 \exp\left(-\frac{J_v \tau}{k}\right)$$

– pores whose fraction f of the section has been obstructed:

$$n_2 = f n_0 \left[1 - \exp\left(-\frac{J_v \tau}{k}\right)\right]$$

This gives us the variation of J_V over time:

$$J_v = J_{v0}\left[f + (1-f)\exp\left(-\frac{J_v \tau}{k}\right)\right]$$

where:

$$0 < f < 1$$

After a sufficient period of time, we obtain:

$$J_v = J_{v0} f$$

This partial blocking is due to the formation of vault structures (or rather domes) within the pores by the accumulation of more than two particles.

From that moment on, the membrane allows the liquid to pass through but stops the particles. We are then dealing with conventional filtration, and the stopped particles form a cake at the surface of the membrane.

For further information, readers can refer to section 1.1.3 which gives the value of the volume of filtrate Ω obtained after a period of time τ. The flowrate of filtrate decreases because the thickness of the filtration cake increases. It must not be forgotten, though, that the advantage of convective filtration is precisely that it prevents the formation of a cake.

2.5.7. *Cleansing of fouled surfaces*

With regard to biofilms, we must draw a distinction between:

– "sanitization", which divides the microbe population by 10^3;

– "disinfection", which should divide this population by 10^6;

– "sterilization", which entirely destroys all microbes.

The cleansing procedures are:

– mechanical curing of the surface of the tubes, by rolling in them balls of foam rubber;

– backwashing (in the opposite direction to normal operation). Backwashing, though, is to be avoided in the case of composite membranes;

– ultrasound waves, which detach the hard deposits (scale).

The chemical agents are:

– caustic soda which, in combination with a detergent, removes silica, organic material and biofilms;

– acids (nitric, acetic, citric acid) which remove scale,

– enzymes which develop, if the pH is near to 7. They destroy the polysaccharides secreted by the microbes, but they must be prevented from attacking the membrane. Cleansing using enzymes is a lengthy process:

– complexing agents such as EDTA;

– disinfectants (oxygenated water and sodium hypochlorite, also known as Javel water).

2.5.8. *Depending on the type of industry*

1) Food industries. Washing first with acid and then with a base may be necessary, because the fouling agents are proteins, salts and sugars. Cleaning is performed daily.

2) Seawater (desalination and cooling of exchangers in the chemical- or petroleum industries). In terms of prevention, gridding, meaning the crossing of a grid, stops the algae, shells and crustaceans. A treatment with chlorine kills marine micro-organisms. An acid- or alkaline treatment may be useful;

3) natural water and wastewater: the situations are very diverse: EDTA (ethylenediaminetetraacetic acid) eliminates calcium from water and can be used for pretreatment. Caustic soda destroys organic substances.

DEPENDING ON THE PORE DIAMETER.–

The frequency N of the cleansing increases with the pore diameter because, *the higher that diameter is, the more easily the pores will clog:*

$$N_{IO} < N_{NF} < N_{UF} < N_{MF}$$

This is why membranes with low permeability clog less quickly.

NOTE.–

A wash temperature of 50°C (which does not deteriorate the membranes) seems to be most effective.

2.5.9. Empirical laws governing fouling

Let J_V represent the liquid flux density expressed in relation to unit surface of the membrane. We can write:

$$J_V = \frac{J_{V0}}{1 + At}$$

or indeed:

$$J_V = J_{V0} \exp(-At)$$

In a slightly less empirical manner, we can add to the resistance R_m of the clean membrane a fictitious resistance R_E expressing the presence of fouling, and we write:

$$J_V = \frac{\Delta P}{\mu(R_m + R_E)}$$

with, for R_E:

– a homographic function:

$$R_E = R_{E_0} \left(\frac{\tau}{\tau + \tau_0} \right)$$

– or an exponential function:

$$R_E = R_{E_0} \left[1 - \exp\left(-\frac{\tau}{\tau_0} \right) \right]$$

– or even an indefinitely-increasing (linear) function:

$$R_E = R_{E_0} \left(\frac{\tau}{\tau_0} \right)$$

– and finally a logarithmic function:

$$R_E = R_{E_0} \ln \left(1 + \frac{\tau}{\tau_0} \right)$$

2.5.10. *Prediction and prevention of fouling*

Only a test in a micropilot with a membrane subject to industrial conditions can ensure accurate prediction of fouling. Indeed, there is no definite match between industrial behavior and tests using a Büchner filter with a membrane, in spite of the existence of multiple and varied "indices".

A good way of preventing the fouling is to increase the velocity of the liquid sweeping the membrane – i.e. the friction of that liquid against the membrane so as to re-entrain the particles which have been deposited.

2.5.11. *Biofilms*

The deposition, adherence and multiplication of the microbes brought by the liquid result in the formation of "biofilms" – i.e. microbial films whose cohesion is ensured by a cement, glycocalyx, which is porous and can therefore allow the passage of nutrients and evacuate waste products.

In order for a biofilm to form, the surface must first be "prepared" by the adsorption of nutrients from the liquid. Hydrophobic membranes readily adsorb biofilms. The thickness of the biofilm is limited by the friction and any potential re-entrainment by the liquid.

Ethylenediaminetetraacetic acid (EDTA) destroys glycocalyx and, thus, the pores of the biofilm are obstructed by the waste products of the glycocalyx and the microbes are no longer fed.

Chlorine is effective but it is preferable to employ it preventatively – i.e. before the appearance of the biofilm.

Theory of Dialysis-Biological Membranes – Electrodialysis

3.1. Introduction

3.1.1. *Means of crossing a membrane separating two solutions at the same pressure*

We distinguish between the following situations:

– the crossing of a membrane by electrically neutral molecules;

– crossing by ions of an electrically charged membrane (negatively charged tomato peel or onion skin);

– crossing of a neutral membrane by ions;

– transfer assisted by a carrier molecule residing in the membrane. This mechanism pertains to ions and also to neutral molecules. A good example is the transfer of oxygen by the carrier, hemoglobin;

– electrodialysis, which is the crossing by ions of a charged membrane under the influence of an external electrical field;

– the crossing of a lipid membrane by living cells through the tunnel effect;

– hemodialysis, which rids blood of its impurities.

However, the above list is not exhaustive. The pH may be important [BAB 80, CUS 71]. Light may affect the transfer, and it does not appear that chlorophyll's transport of oxygen and carbon dioxide under the influence of light has been sufficiently investigated.

3.1.2. *Liquid membranes*

Researchers sometimes create what they call "liquid membranes" (which have no industrial applications at present). The method is as follows:

– rapid agitation to form an emulsion of water in oil, and this emulsion is stabilized by a surfactant;

– this emulsion is added to a second aqueous phase. Then, moderate agitation is applied. The result is a water-in-oil-in-water emulsion.

The oily phase is the liquid membrane separating the two aqueous phases. An example of this is given in [HOC 75, DUF 78] and [LEE 78].

The easy construction of liquid membranes means they can be used to study certain transfer mechanisms.

3.2. Electrically neutral solutes

3.2.1. *The Maxwell–Stefan equation in a membrane*

Outside of a pore, Fick's law is written:

$$c_0 \frac{d\mu}{dz_0} = \frac{RT}{D_0} J_0$$

If z_m is the thickness of the membrane and t is the tortuosity:

$$z_0 = t z_m$$

The flux density of the solute is such that:

$$\frac{J_0}{J_m} = \frac{t}{\varepsilon}$$

Indeed:

– in pores, the solute must, in the same time period, cover a distance t times greater than if the pores were perpendicular to the face of the membrane;

– with respect to the face of the membrane, the flux density J_m is ε times what it is in relation to the section of the pores.

Finally, the concentration c_0 in the liquid of the pores is equal to the concentration c_m in relation to the volume of the membrane divided by the porosity ε.

$$\frac{c_0}{c_m} = \frac{1}{\varepsilon}$$

Fick's equation then becomes:

$$\frac{c_m}{\varepsilon}\frac{d\mu}{dz_m} \times \frac{1}{t} = \frac{RT}{D_0}J_m\frac{t}{\varepsilon}$$

On condition that we set:

$$D = \frac{D_0}{t^2}$$

Fick's equation is written:

$$c_m\frac{d\mu}{dz} = \frac{RT}{D}J_m$$

Similarly, using the same relation $(D = D_0/t^2)$ for the diffusivities, the Maxwell–Stefan equation is written:

$$-c_i\frac{d\mu_i}{dz} = RT\sum_{j=1}^{n}\left(\frac{J_i x_j - J_j x_i}{D_{ij}}\right)$$

z: dimension perpendicular to the membrane (m)

J: molar flux density in relation to the frontal surface of the membrane ($kmol.m^{-2}.s$)

c_i: concentration expressed in relation to the overall volume of the membrane

$$D_{ij} = \frac{D_{0ij}}{t^2}$$

3.2.2. Influence of the molecular diameter

In reality, the diffusivity D_{ij} must be obtained by:

$$D_{ij} = \frac{D_{0ij} K_d}{t^2}$$

The coefficient K_d is obtained as a function of the ratio:

$$\lambda = \frac{\text{radius of the solute}}{\text{mean pore radius}} = \frac{r_s}{r_p}$$

$0 < \lambda \le 0.8$ $\qquad\qquad$ $K_d = 1 - 2.30\lambda + 1.154\lambda^2 + 0.224\lambda^3$

$0.8 < \lambda \le 1$ $\qquad\qquad$ $K_d = -0.105 + 0.318\lambda - 0.213\lambda^2$

These coefficients are cited by Bandini and Vezzani [BAN 03].

The molecular radius r_s of the solute can be obtained on the basis of knowledge of the solute's diffusivity in the solution.

$$r_s = \frac{kT}{6\pi\mu D}$$

μ: viscosity of the solution (Pa.s)

k: Boltzmann's constant (1.38032×10^{-23} $J.K^{-1}$)

T: absolute temperature (K)

It can also be obtained by evaluating the volume of a supposedly spherical molecule:

$$\frac{V_m}{N_A} = \frac{4\pi r_p^3}{3} = \frac{M}{\rho_L N_A} \text{ and therefore } r_s = \left(\frac{3M}{4\pi\rho_L N_A}\right)^{1/3} = \left(\frac{3V_m}{4\pi N_A}\right)^{1/3}$$

M: molar mass (kg.kmol^{-1})

ρ_L: density of the pure solute in the liquid- or solid state (kg.m^{-3})

N_A: Avogadro's number: 6.023×10^{26} molecules per kmol

V_m: molar volume (m^3.kmol^{-1})

3.2.3. *Crossing of the membrane*

Suppose we have a membrane separating two solutions of differing compositions, at the same pressure and the same temperature.

If e_m is the thickness of the membrane, the chemical-potential gradient will be written as follows, between solutions A and B separated by the membrane:

$$\frac{d\mu_i}{dz} = \frac{RT}{e_m}\left(Ln\left(\gamma_{Ai}c_{Ai}\right) - Ln\left(\gamma_{Bi}c_{Bi}\right)\right)$$

Often, the concentrations are sufficiently low for Fick's law to be sufficient, without the need to use the whole of the Maxwell–Stefan law.

3.3. Electrically charged membrane: conventions

3.3.1. *Preliminary hypotheses*

Hereafter, we consider only the concentration of the ions in the pores or, more generally, in the liquid phase responsible for the swelling of the membrane.

The hypothesis whereby there is no liquid flowrate and the permeation of the ions takes place solely under the influence of the gradients of chemical potential and of electrical potential corresponds to what happens in the operation of dialysis, and, in particular, characterizes exchanges across the membrane of living biological cells.

However, let it be stated right now that the crossing of a biological membrane also takes place very often by a phenomenon known as "assisted permeation", which we shall examine in section 3.6.

3.3.2. Definitions

The fixed charge X of a membrane, expressed in Faraday, can be viewed in relation to the overall volume of the swollen membrane or in relation only to the volume of the impregnating liquid.

$$X_m = \frac{\text{fixed charge}}{\text{total volume of the membrane}} \qquad \left(F.m^{-3} \right)$$

$$X_0 = \frac{\text{fixed charge}}{\text{volume of impregnating liquid}} \qquad \left(F.m^{-3} \right)$$

$$X_m = \varepsilon X_0$$

ε: porosity of the membrane (liquid fraction in terms of volume).

In general, the value X is preceded by a coefficient σ equal to +1 for positive charges and −1 for negative charges.

The Donnan equilibrium can be used to calculate the equilibrium between the concentration of a solution outside of the membrane and the concentration of the impregnating liquid.

In our discussion here, we shall set:

$$X = X_0$$

PRELIMINARY NOTE.–

Unlike we did for neutral molecules, in the case of ions, we shall not take their size into account. Generally speaking, the ions which are of interest to a physicist specializing in membranes are small.

3.4. Ionic concentrations inside and outside the membrane

3.4.1. *Donnan equilibrium (ions of whatever valence)*

Consider two phases α and β, at equilibrium and separated by an interface which is permeable to the ions and to the solvent. The equilibrium results in the equality of the electrochemical potentials. Thus, for the ion i whose valence is z_i:

$$z_i kFF\Phi_\alpha + RT\ln a_{i\alpha} = z_i F\Phi_\beta + RT\ln a_{i\beta}$$

F: Faraday

or indeed:

$$Ln\frac{a_{i\alpha}}{a_{i\beta}} = \frac{kF}{RT}\left(\phi_\beta - \phi_\alpha\right)z_i = z_i Lnk_{\beta\alpha}^{zi} \qquad [3.1]$$

or else:

$$a_{i\alpha} = a_{i\beta}k_{\beta\alpha}^{zi} \quad \text{or indeed} \quad c_{i\alpha} = c_{i\beta}\frac{\gamma_{i\beta}}{\gamma_{i\alpha}}k_{\beta\alpha}^{zi} = c_{i\beta}\Gamma_i k_{\beta\alpha}^{zi}$$

Let us write the electrical neutrality of the α phase, which is now supposed to be the phase interior to the membrane of charge σXF in the pores:

$$\sigma XF + \sum_i c_{i\alpha}kFz_i = 0 \quad \text{or indeed} \quad \sigma X + \sum_i c_{i\alpha}z_i = 0$$

kF: Faraday: 96.487×10^6 C.keq^{-1}

However, the concentrations $c_{i\alpha}$ are unknowns, and it is the concentrations outside of the membrane $c_{i\beta}$ that are known. Thus, we write, for the neutrality of the internal α phase:

$$\sigma X + \sum_i c_{i\beta} \Gamma_i z_i k_{\beta\alpha}^{zi} = 0$$

Strictly speaking, when we know the β phase, we know the $\gamma_{i\beta}$ but not the $\gamma_{i\alpha}$. Thus, to begin with, we assimilate all the activities to the concentrations and write the following, as $\gamma_{i\beta} = \gamma_{i\alpha} = 1$.

$$\sigma X + \sum_i c_{i\beta} z_i k_{\beta\alpha}^{zi} = 0 \qquad \text{when} \qquad c_{i\alpha} = c_{i\beta} k_{\beta\alpha}^{zi}$$

This is an algebraic equation in terms of $k_{\beta\alpha}$ whose positive roots we shall retain and, of those positive roots, we use those which are such that the concentration of the ions of the same sign as σX is less than that of the ions of the opposite sign to σX. More specifically, we are dealing with:

– counter-ions if σX and z_i are of opposite signs;

– co-ions if σX and z_i are of the same sign.

For the sake of conciseness, we call the counter-ions g-ions (to recap, the g comes from the German *gegen*, meaning "counter"), and the co-ions are c-ions.

When we know this initial estimation of the $c_{i\beta}$, we can calculate the $\gamma_{i\beta}$ and thus the Γ_i. The algebraic equation in $k_{\beta\alpha}$ then becomes:

$$\sigma X + \sum_i c_{i\beta} \Gamma_i z_i k_{\beta\alpha}^{zi} = 0$$

This gives us a new estimation of the $c_{i\beta}$ values. Two or three iterations will suffice.

3.4.2. *Donnan exclusion*

The above means that the c-ions are excluded from the pores to a greater or lesser extent, whilst the g-ions are driven into the pores – particularly if they are multivalent. Similarly, a c-ion is rejected more strongly when it is multivalent.

This situation is what characterizes the Donnan exclusion.

3.4.3. *Total potential different and Donnan potentials*

What we measure is the total potential φ_T, which is the difference between the external potentials: $\varphi_{E2} - \varphi_{E1}$.

We can write:

$$\varphi_T = \varphi_{E2} - \varphi_{I2} + \varphi_{I2} - \varphi_{I1} + \varphi_{I1} - \varphi_{E1}$$

We know that the Donnan potential is (see equation [3.1]):

$$\varphi_D = \varphi_E - \varphi_I$$

Hence:

$$\varphi_T = \left(\varphi_{I2} - \varphi_{I1}\right) + \left(\varphi_{D2} - \varphi_{D1}\right)$$

By calculating the φ_D and having measured φ_T, we can deduce the internal potential of the membrane:

$$\varphi_m = \varphi_{I2} - \varphi_{I1}$$

3.4.4. *Case study (monovalent ions)*

Let k represent the sharing coefficient between, firstly, the concentrations c_{iI}^+ of the cations and c_{jI}^- of the anions in the liquid of the membrane and, secondly, the concentrations c_{iE}^+ and c_{jE}^- in the external solution.

$$k = c_{iI}^+ \Big/ c_{iE}^+ \qquad \frac{1}{k} = c_{jI}^- \Big/ c_{jE}^-$$

The neutrality of the liquid phase of the membrane is written as follows, when $z_i = 1$ and $z_j = -1$.

$$\sigma X + \sum_i k c_{iE}^+ - \sum_j \frac{c_{jE}^-}{k} = 0$$

Let us set:

$$\sum_i c_{iE}^+ = c_E^+ \quad \text{and} \quad \sum_j c_{jE}^- = c_E^- \quad \text{with} \quad c_E^+ = c_E^- = c_E$$

The equation in terms of k is written:

$$k^2 + \frac{\sigma X}{c_E^+} k - 1 = 0$$

The roots of this equation are:

$$k = \sqrt{1 + \frac{X^2}{\left(2c_E^+\right)^2}} - \frac{\sigma X}{2c_E^+} \quad \text{and} \quad \frac{1}{k} = \sqrt{1 + \frac{X^2}{\left(2c_E^+\right)^2}} + \frac{\sigma X}{2c_E^+}$$

In addition, we can see that, if $\sigma = 1$

$$k < 1/k$$

Put differently, in the membrane, we can verify that the concentration of the c-ions is less than that of the g-ions.

The Donnan potential is:

$$\varphi_D = \varphi_E - \varphi_I = \frac{RT}{F} Ln\, k$$

We set:

$$\sqrt{4c_E^2 + X^2} = x \quad \text{and thus} \quad c_E^2 = \frac{x^2 - X^2}{4}$$

$$k = \frac{x - \sigma X}{2c_E^+}$$

Note that, symmetrically, we can write:

$$\varphi_D = \frac{RT}{2kF}\ln\frac{(x-\sigma X)^2}{(2c_E^e)^2} = \frac{RT}{2kF}Ln\frac{(x-\sigma X)^2}{(x^2-(\sigma X))^2} = \frac{RT}{2kF}Ln\frac{(x-\sigma X)}{(x+\sigma X)}$$

3.5. Internal potential and concentrations

3.5.1. *Calculation of the internal potential: ions all monovalent* *[TEO 51]*

The transport of cations in the membrane results in the transport:

$$J_i^+ = \frac{-RT}{F}u_i c_i^+\left[\frac{dLnc_i^+}{dx}+\frac{F}{RT}\frac{d\varphi}{dx}\right]$$

We have assimilated the activity of the ions to their concentration. This transport equation contains a diffusion term and an electric migration term.

For a monovalent cation, we indeed have:

$$\frac{RT}{kF}u_i^+ = D_i$$

u_i: mobility of the cation i ($m^2.s^{-1}.V^{-1}$)

Let us set:

v_j: absolute value of the mobility of the anion j ($m^2.s^{-1}.V^{-1}$)

φ: electrical potential (V):

$$\frac{kF\varphi}{RT} = Ln\xi \qquad\qquad\qquad [3.1bis]$$

F: Faraday's constant: 96.487×10^6 $C.keq^{-1}$

R: ideal gas constant: 8314.39 $J.kmol^{-1}.K^{-1}$

The transport equations for the cations and anions become:

$$J_i^+ = \frac{-RT}{F} u_i c_i^+ \left[\frac{d \operatorname{Ln} c_i^+}{dx} + \frac{d \operatorname{Ln} \xi}{dx} \right] = -u_i A_i \qquad [3.2]$$

$$J_j^- = \frac{-RT}{F} v_j c_j^- \left[\frac{d \operatorname{Ln} c_j^-}{dx} - \frac{d \operatorname{Ln} \xi}{dx} \right] = -v_j B_j \qquad [3.3]$$

We set:

$$A = \sum_i A_i \quad \text{and} \quad B = \sum_i B_i$$

$$c^+ = \sum_i c_i^+ \qquad c^- = \sum_j c_j^-$$

As the fluxes are constant, the parameters A_i, B_j, A and B are constant throughout the thickness of the membrane, but the ratio A/B remains unknown.

The neutrality equation is written:

$$c^+ - c^- + \sigma X = 0$$

Thus:

$$c^+ + c^- = 2 c^+ + \sigma X$$

Let us form the combinations:

$$A + B = \frac{RT}{kF} \frac{d\left(c^+ + c^-\right)}{dx} + RT\left(c^+ - c^-\right) \frac{d \operatorname{Ln} \xi}{dx}$$

or indeed:

$$A + B = \frac{2RT}{kF} \frac{dc^*}{dx} - \frac{RT}{F} \sigma X \frac{d \operatorname{Ln} \xi}{dx} \qquad [3.4]$$

Similarly:

$$A - B = \frac{RT}{kF}\left(2c^{+} + \sigma X\right)\frac{d \operatorname{Ln} \xi}{dx} \qquad [3.5]$$

Let us eliminate dx between equations [3.4] and [3.5]. We find:

$$c^{+} + \frac{A\sigma X}{A+B} = \frac{A-B}{A+B}\frac{dc^{+}}{d \operatorname{Ln} \xi} \qquad [3.6]$$

By integrating between faces 1 and 2 of the membrane:

$$\ln \xi = \frac{A-B}{A+B} \operatorname{Ln}\left[\frac{c_2^{+} + \dfrac{A}{A+B}\sigma X}{c_1^{+} + \dfrac{A}{A+B}\sigma X}\right] \qquad [3.7]$$

We then obtain the expression of the internal potential φ (see equation [3.1b]):

$$\frac{F\varphi}{RT} = \operatorname{Ln}\xi = q \operatorname{Ln}k \qquad [3.8]$$

where:

$$q = \frac{A-B}{A+B}\left(\text{hence } \frac{A}{A+B} = \frac{1+q}{2} = \frac{\operatorname{Ln}k\xi}{2\operatorname{Ln}k}\right) \qquad [3.9]$$

and:

$$k = \frac{c_2^{+} + \dfrac{A}{A+B}\sigma X}{c_1^{+} + \dfrac{A}{A+B}\sigma X} = \frac{c_2^{+} + \dfrac{1+q}{2}\sigma X}{c_1^{+} + \dfrac{1+q}{2}\sigma X} \qquad [3.10]$$

3.5.2. *Flux density of each ion*

Let us begin by integrating equation [3.4]:

$$A + B = \frac{2RT}{e_m kF}\left[c_2^+ - c_1^+ - 0.5\sigma X \ln \xi\right] \qquad [3.11]$$

In view of equations [3.8] and [3.9]:

$$\frac{A}{A + B} = \frac{\ln(k\xi)}{2\ln k} \qquad [3.12]$$

Teorell, however, made the following hypothesis:

$$\frac{A_i}{A} = \frac{c_{2i}^+ \xi - c_{1i}^+}{c_2^+ \xi - c_1^+} \qquad [3.13]$$

Therefore, we can write:

$$J_i^+ = -u_i A_i = -u_i \frac{A_i}{A} \times \frac{A}{(A+B)} \times (A+B)$$

Put differently:

$$J_i^+ = -\frac{u_i RT}{e_m kF}\left(c_2^+ - c_1^+ - 0.5\sigma \times \ln\xi\right)\frac{Lnk\xi}{Lnk} \times \frac{c_{2i}^+ \xi - c_{1i}}{c_2^+ \xi - c_1^+} \qquad [3.14]$$

For anions, we need to change c^+ and c_i^+ to c^- and c_j^-, u to v and ξ to $1/\xi$, with the exception of the contents of the parenthesis, which is not altered.

3.5.3. *Electrical neutrality of the total ionic flux*

In this case, the total flux of cations is equal to the total flux of anions.

We have established that (see equation [3.14]):

$$J_i^+ = -u_i \left[\frac{RT}{e_m F}\frac{\left(c_2^+ - c_1^+ - 0.5\sigma \times \ln\xi\right)}{\ln k}\right]\ln(k\xi) \times \frac{c_{i2}^+ \xi - c_{i1}^+}{c_2^+ \xi - c_1^+}$$

or indeed:

$$J_i^+ = -u_i K \operatorname{Ln}(k\xi) \times \frac{c_{i2}^+\xi - c_{i1}^+}{c_2^+\xi - c_1^+}$$

and:

$$J^+ = \sum_j J_j^+ = \frac{K\operatorname{Ln}(k\xi)}{c_2^+\xi - c_1^+}\left[\sum_j u_i c_{i2}^+\xi - \sum_i u_i c_{i1}\right]$$

$$J^- = \sum_j J_j^- = \frac{K\operatorname{Ln}(k/\xi)}{c_2^- - c_1^-\xi}\left[\sum_j v_j c_{j2}^- - \sum_j v_j c_{j1}^-\xi\right]$$

By writing that $J^+/J^- = 1$ and setting:

$$U = \sum_i u_i c_i^+ \quad \text{and} \quad V = \sum_j v_i c_i^-$$

we obtain:

$$\frac{U_2\xi - U_1}{V_2 - V_1\xi} = \frac{c_2^+\xi - c_1^+}{c_2^- - c_1^-\xi} \times \frac{\operatorname{Ln}(k/\xi)}{\operatorname{Ln}(k\xi)} \qquad [3.15]$$

This relation was established by [PLE 30, p. 743] without using Teorell's hypothesis (equation [3.13]).

3.5.4. *Isotonic solutions*

Two solutions are said to be isotonic if their overall concentrations are equal and if, of course, each of them is electrically neutral. We can also say that the activity of the solvent is the same in the two solutions. In this situation, the system of equations [3.10] shows that:

$$k = 1$$

Thus, because $c_2^+ = c_1^+ = c_2^- = c_1^-$, equation [3.15] becomes:

$$\frac{U_2\xi - U_1}{V_2 - V_1\xi} = -1 \text{ or indeed } \xi = \frac{U_1 - V_2}{U_2 - V_1}$$

For the membrane separating the cytoplasm of a cell and an external solution (blood serum, plant sap), these solutions will be isotonic as a general rule.

3.5.5. *Practical calculation of the internal potential and of the ionic fluxes*

We take the c_{1i}^+ and c_{2i}^+ (thus the c_1^+ and c_2^+) and:

$$q = \frac{A - B}{A + B}, \text{ meaning that } \frac{A}{A + B} = \frac{1+q}{2} \qquad [3.9]$$

According to equation [3.10]:

$$k = \frac{c_2^+ + \dfrac{1+q}{2}\sigma X}{c_1^+ + \dfrac{1+q}{2}\sigma X}$$

According to equation [3.8]:

$$\text{Ln}\,\xi = q\,\text{Ln}\,k$$

For each value of q, there is a corresponding value of ξ associated with a value of each of the two sides of equation [3.15]. We need to find the value of q such that the two sides are equal. Thus, we obtain the value of ξ, and hence the potential φ.

The ionic fluxes can be deduced from this by means of equation [3.14].

3.5.6. *Profiles of the internal potential and concentrations*

We know ξ, c_1^+, c_2^+ and q, which we determined in section 3.5.5, and we need to calculate ξ_x.

After integration of equation [3.4], we obtain:

$$\frac{x}{e_m} = \frac{c_x^+ - c_1^+ - 0,5\sigma X \operatorname{Ln}\xi_x}{c_2^+ - c_+^1 - 0,5\sigma X \operatorname{Ln}\xi_x} \qquad [3.16]$$

Equation [3.17] is obtained on the basis of equation [3.7]:

$$\operatorname{Ln}\xi_x = q \operatorname{Ln}\left[\frac{c_x^+ + \dfrac{1+q}{2}\sigma X}{c_1^+ + \dfrac{1+q}{2}\sigma X}\right] \qquad [3.17]$$

By successive approximations, these two equations enable us to calculate $\operatorname{Ln}\xi_x$, and c_x^+ corresponding to the distance x from the inlet. The calculation of k_x can be deduced from this.

If we stop at the slice of abscissa x before face 2, equation [3.14] is written:

$$J_i^+ = -u_i \frac{RT}{Fx}\left(c_x^+ - c_1^+ - 0,5\sigma X \operatorname{ln}\xi_x\right)\frac{\operatorname{Ln}k_x\xi_x}{\operatorname{Ln}k_x} \times \frac{c_{xi}^+\xi_x - c_{1i}}{c_x^+\xi_x - c_1^+} \qquad [3.18]$$

Let us divide equations [3.14] and [3.18], term by term, in light of equation [3.16]:

$$1 = \frac{\left(c_{xi}^+\xi_x - c_{1i}^+\right)\left(c_2^+\xi_2 - c_1\right)}{\left(c_x^+\xi_x - c_1^+\right)\left(c_{2i} - c_{1i}\right)} \times \frac{\operatorname{Ln}k_x\xi_x}{\operatorname{Ln}k\xi} \times \frac{\operatorname{Ln}k}{\operatorname{Ln}k_x}$$

This relation gives c_{xi} as a function of x.

Teorell's calculations show that, if the system of solutions contains three different types of ions, the concentration of one of the ions may exhibit a maximum in the thickness of the membrane. Teorell confirmed this experimentally with a membrane composed of multiple superposed sheets, analyzed separately.

This phenomenon corresponds to what happens at the instantaneous junction of two solutions of different compositions, studied by [PLE 30]. The two solutions were not immiscible (Figure 1 in Plettig's communication).

NOTE.–

Schlögel [SCH 54] proposed a generalization of Teorell's theory for electrolytic solutions of any composition.

3.5.7. *Transport of a single compound [MEY 36]*

Let us write the *neutrality equation* for the free migration, simplifying by dx.

$$J^+ dx = J^- dx \quad \text{or} \quad uc^+ \left[dLnc^+ + \frac{kF}{RT} d\phi \right] = vc^- \left[dLnc^- - \frac{kF}{RT} d\phi \right]$$

Meyer and Sievers [MEY 36] set:

$$c^+ = \frac{x - \sigma X}{2} \text{ and } c^- = \frac{x + \sigma X}{2}, \text{ where } x = \sqrt{(\sigma X)^2 + 4c_E^{+2}}$$

The neutrality equation becomes:

$$\frac{u - v}{u + v} \times \frac{dx}{x - \sigma X \left(\dfrac{u - v}{u + v} \right)} = -\frac{kF}{RT} d\phi$$

The internal potential in the membrane is then:

$$\phi_2 - \phi_1 = \phi = \frac{u - v}{u + v} \frac{RT}{F} Ln \left[\frac{x_1 - \sigma X \dfrac{u - v}{u + v}}{x_2 - \sigma X \dfrac{u - v}{u + v}} \right]$$

where:

$$x_1 = \sqrt{X^2 + 4c_{E1}^{+2}} \quad \text{and} \quad x_2 = \sqrt{X^2 + 4c_{E2}^{+2}}$$

Note that if $x_1 > x_2$, then $\varphi_2 > \varphi_1$. Experience confirms this result.

NOTE.–

Both Teorell and Meyer & Sievers find the following form for the internal potential.

$$\varphi = \alpha Ln \frac{c_1^+ + \beta}{c_2^+ + \beta}$$

This expression has been verified experimentally by Meyer and Sievers.

NOTE.–

Non-charged membrane crossed by ions.

Teorell's results, as well as Meyer and Sievers', remain valid on condition that we set:

$$X \equiv 0$$

3.5.8. *Numerical integration of the diffusion/migration equations*

For the ion i, the diffusion/migration equation is:

$$J_i = -\frac{D_i}{RT}\frac{dc_i}{dx} - c_i u_i \frac{d\varphi}{dx}$$

c_i: concentration of the ion (kion.m^{-3})

u_i: algebraic mobility of the ion i (m^2.s^{-1}.V^{-1})

φ: internal potential of the membrane (V)

Let us write the electrical neutrality of the flux crossing the membrane:

$$\sum_i z_i J_i = 0$$

Thus:

$$\frac{d\varphi}{dx} = -\frac{\displaystyle\sum_i \frac{D_i z_i}{RT}\frac{dc_i}{dx}}{\displaystyle\sum_i c_i u_i z_i}$$

z_i: algebraic valence of the ion i

By feeding this expression of $d\varphi/dx$ in all the diffusion/migration equations, we see that the J_i are linear forms of the dc_i/dx. We can therefore resolve this system as a function of the dc_i/dx and integrate using the Runge–Kutta method (see Appendix 1), provided we know the J_i. However, this is what we are looking for.

Let us assign the indices 0 and n to the internal concentrations on the faces of the membrane. These concentrations are givens in the problem:

$$c_{0i}^{(0)} \text{ and } c_{ni}^{(0)}$$

Let us take an initial estimation of the J_i, which are of the form:

$$J_i^{(0)} = F\left((dc_j/dx)^{(0)},...,(dc_j/dx)^{(0)},...,(dc_k/dx)^{(0)}\right)$$

We set:

$$\left(\frac{dc_j}{dx}\right)^{(0)} \# \frac{c_{0j}^{(0)} - c_{nj}^{(0)}}{e_m}$$

The Runge–Kutta integration over the thickness of the membrane gives the values of the $c_{ni}^{(1)}$. We set:

$$1 + \eta_1 = \frac{c_{0i}^{(0)} - c_{ni}^{(1)}}{c_{0i}^{(0)} - c_{ni}^{(0)}}$$

We then correct J_i by setting:

$$J_i^{(1)} = J_i^{(0)} \left(1 - \frac{\eta_1}{2}\right)$$

and, more generally:

$$J_i^{(m)} = J_i^{(m-1)} \left(1 - \frac{\eta_m}{2}\right)$$

where:

$$1 + \eta_m = \frac{c_{0i}^{(0)} - c_{ni}^{(m-1)}}{c_{0i}^{(0)} - c_{ni}^{(0)}}$$

Three or four iterations will suffice.

Thus, having converged on the profiles of c_i, we can calculate the profile of the internal potential φ.

3.6. Assisted permeation

3.6.1. *Definition*

We also speak of "facilitated diffusion". The compound A which needs to pass through the membrane is combined with a chemical species B, known as the support, which does not leave the membrane. The combination AB passes through the membrane and dissociates on exiting the membrane, liberating A. The support B makes the return journey alone, and when it reaches the inlet, it combines with a new molecule of A.

Thus, the support B acts as a shuttle between the two faces of the membrane.

3.6.2. *General*

The reaction of combination of the transported product A with the support B is:

$$A + B \underset{k_2}{\overset{k_1}{\rightleftharpoons}} AB$$

In the thickness of the membrane, the balances of the compounds A, B, and AB are written thus:

$$D_A \frac{d^2 c_A}{dx^2} = k_1 c_A c_B - k_2 c_{AB} \qquad [3.19]$$

$$D_B \frac{d^2 c_B}{dx^2} = k_1 c_A c_B - k_2 c_{AB} \qquad [3.20]$$

$$D_{AB} \frac{d^2 c_{AB}}{dx^2} = k_1 c_A c_B + k_2 c_{AB} \qquad [3.21]$$

Let us subtract equation [3.20] from equation [3.19], term by term, and integrate twice:

$$D_A c_A - D_B c_B = a_1 x + a_2 \qquad [3.22]$$

We then add equations [3.19] and [3.21] together, term by term, and integrate twice:

$$D_A c_A + D_{AB} c_{AB} = a_3 x + a_4 \qquad [3.23]$$

Next, we add equations [3.20] and [3.21] together, term by term, and integrate twice:

$$D_B c_B + D_{AB} c_{AB} = a_5 x + a_6 \qquad [3.24]$$

However, if we suppose that the support B does not exit or enter the membrane:

$$D_B \frac{dc_B}{dx} + D_{AB} \frac{dc_{AB}}{dx} = 0 \quad \text{for} \quad x = 0 \quad \text{and also} \quad x = e_m$$

Thus:

$$a_5 = 0 \qquad [3.25]$$

Let us write that [3.23]–[3.24] = [3.22]

$$a_3x + (a_4 - a_6) \equiv a_1x + a_2$$

Consequently:

$$a_3 = a_1 \quad \text{and} \quad a_6 = a_4 - a_2$$

However, equation [3.24] is simply a consequence of equations [3.22] and [3.23]. It is therefore more helpful to replace equation [3.24] with equation [3.19].

The system of equations then becomes:

$$D_A c_A - D_B c_B = a_1x + a_2 \qquad\qquad [3.22]$$

$$D_A c_A - D_{AB} c_{AB} = a_1x + a_4 \qquad\qquad [3.23]$$

$$\frac{d^2 c_A}{dx^2} = \frac{k_1}{D_A} c_A c_B - \frac{k_2}{D_A} c_{AB} \qquad\qquad [3.19]$$

By replacing c_B and c_{AB} in equation [3.19] with their expressions drawn from equations [3.22] and [3.23], we obtain the following by setting $c_A = y$:

$$y'' = y^2 \frac{k_1}{D_B} - yx\frac{k_1 a_1}{D_A D_B} + y\left(\frac{k_2}{D_{AB}} - \frac{k_1 a_2}{D_A D_B}\right) - x\frac{a_1 k_2}{D_A D_{AB}} - \frac{k_2 a_4}{D_A D_{AB}}$$

The parameters a_1, a_2 are defined by the equations (the parameter a_6 is supposed to be a given).

$$y(0) = c_A(0)$$
$$y(e_m) = c_A(L)$$

Ward [WAR 70] indicates the proper method to solve this differential equation. This is the Galerkin method [NOU 85, p. 276]. The function y and the corrective function $\hat{y}$ are expressed by truncated series of Chebyshev polynomials [NOU 85, p. 276]. First, however, we must set:

$$z = \frac{2x - e_m}{e_m} \quad \text{meaning that } x = \frac{e_m(z+1)}{2}$$

when:

$$x = 0 \qquad z = -1$$

$$x = e_m \qquad z = 1$$

e_m is the thickness of the membrane.

Ward [WAR 70] gives certain specifications for the use of this method. It seems that the finite-difference method is difficult to apply to this problem, owing to the presence of the term in y^2.

Ward [WAR 70] explains how this calculation method can be used to express the assisted transport of oxygen by hemoglobin.

3.6.3. *Case study I: the supporting reaction is at equilibrium at all points*

This occurs if the coefficients k_1 and k_2 are sufficiently high. The equilibrium is expressed by:

$$c_{AB} = K c_A c_B \qquad [3.26]$$

Very generally, the support B is a large molecule, whereas A is a small molecule, so the volumes of B and AB are practically identical. Consequently, we can set:

$$D_{AB} = D_B = D$$

Equation [3.24], in light of equation [3.25], becomes:

$$c_B + c_{AB} = \text{const.} = \overline{c} \qquad [3.27]$$

We now eliminate c_B between equations [3.26] and [3.27]. We find:

$$c_{AB} = \frac{K \overline{c} c_A}{1 + K c_A}$$

By applying Fick's equation to A and to AB, we obtain the flux density N_A of the component A:

$$N_A = -D\frac{dc_A}{dx} - D\frac{dc_{AB}}{dx} = -D\left[\frac{dc_A}{dx} + \frac{d}{dx}\left(\frac{K\bar{c}c_A}{1+Kc_A}\right)\right]$$

Let us integrate for x varying from 0 to e_m – i.e. from the α phase to the ω phase, taking account of the equilibrium equations.

$$c_{Ao} = Hc_\alpha \quad \text{and} \quad c_{AL} = Hc_\omega$$

We obtain:

$$N_A = \frac{DH}{e_m}(c_\alpha - c_\omega) - \frac{DHK\bar{c}}{e_m}\left(\frac{c_\omega}{1+KHc_\omega} - \frac{c_\alpha}{1+KHc_\alpha}\right)$$

$$N_A = \frac{DH}{e_m}(c_\alpha - c_\omega)\left[1 + \frac{K\bar{c}}{(1+KHc_\omega)(1+KHc_\alpha)}\right]$$

Note that N_A does not depend on x. In other words, what goes into the membrane is equal to what comes out of it. The concentration $\bar{c}$ is a given in the problem.

3.6.4. Case study II: the diffusions of B and AB are quasi-instantaneous in comparison to the reaction rate

It stems from this hypothesis that, across the thickness of the membrane, the concentrations of B and of AB are constant. Thus, it results that equation [3.19] is a second-order linear differential equation that is easy to integrate.

The solution to this equation is to be found in Ward [WAR 70].

To calculate c_B and c_{AB}, we use:

– equation [3.27] which, this time, does not stem from a reasoning process, and simply expresses that the quantity of support B in the membrane is a given in the problem (if we make $D_B = D_{AB} = D$),

– the equality of the flux densities of A on entering and exiting the membrane.

$$\frac{dc_A}{dx}\bigg|_o = \frac{dc_A}{dx}\bigg|_{e_m} \qquad\qquad [3.28]$$

This equation is an irrational equation which contains exponentials and can only be solved by successive approximations. To find an initial approximation of c_B, we suppose, as Ward [WAR 70] does, that the concentration profile of c_A has the shape shown in Figure 3.1.

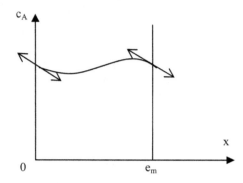

Figure 3.1. *Supposed concentration of A in the thickness of the membrane*

Ward then writes:

$$\frac{d^2 c_A}{dx^2}\bigg|_o = -\frac{d^2 c_A}{dx^2}\bigg|_L$$

This equation is simpler to solve than equation [3.28], but contrary to Ward's claim, this equation does not solve the problem, but merely gives us an initial approximation.

3.7. Permeation by the tunnel effect

3.7.1. *Principle*

This method is also known as "pore transport". Certain large molecules have the form of hollow tubes (gramicidin, alamethicin) which, placed end to end, create a genuine tunnel whose external wall is lipophilic and whose internal wall is hydrophilic.

Thus, cations, which are easily hydrated, can easily pass through a lipid membrane. The permeability of such a membrane is a decreasing function of the size of the ion. Thus, for the permeability:

$$H^+ > K^+ > Na^+ > Li^+$$

Tunnel permeation was described by Haydon and Hadky [HAY 72].

3.8. Hemodialysis

3.8.1. *Definition*

Hemodialysis brings blood and a solution into contact across a membrane which was, formerly, made of cellulose but, today, is made of cuprophane (which is a thin film of cellulose regenerated by the cuprammonium process).

The blood and the solution which is to extract the impurities from the blood (urea, uric acid, creatinine and other toxins generated by the metabolism of living cells) are almost isotonic, meaning that their ionic strengths are similar. Thus, there is only very little water transferred from the blood into the solution. The electrolyte previously dissolved in the extractive solution is sodium chloride for the most part.

The blood contains heparin to prevent coagulation on contact with the membrane and especially with the tubing system. Cellulose, though, is largely biocompatible.

Medical professionals speak of the "clearance", which is the rate at which toxins are transferred across the membrane.

The membranes may be flat (and coiled up in a spiral) or else take the form of hollow fibers.

3.8.2. *Material transfer*

The extraction of the neutral molecules must overcome three types of resistance:

– the limiting laminar layer of blood;

– the resistance of the membrane;

– the limiting laminar layer of the extraction solution.

We accept that the crossing of each resistance is proportional to the difference between the upstream and downstream concentrations.

If the impurities to be extracted are ionic, we can refer to Teorell's theory (see section 3.5).

Hemodialyzers have a membrane surface of 0.2 m^2 to 1.5 m^2, and the thickness of the membrane varies from 10 μm to 30 μm [BRU 89]. The pressure drop on the side of the blood is 10–20 mmHg, because the blood flowrate is between 10 and 50 mL.mn^{-1}. There is a difference in pressure for the transfer of a small amount of water.

$$\Delta P = P_{blood} - P_{solut} \text{ where } 50 \text{ mmHg} < \Delta P < 200 \text{ mmHg}$$

3.9. Electrodialysis

3.9.1. *Principle*

Electrodialysis is based on the properties of ion-exchange membranes – i.e. membranes contain a fixed charge, which may be positive or negative depending on the membrane. The absolute value of that fixed charge varies from 10^{-3} kequiv.kg^{-1} to 5×10^{-3} kequiv.kg^{-1}.

The impregnating liquor of a charged membrane is poor in c-ions (co-ions), of the same sign as that of the fixed charge of the membrane. On the other hand, the concentration in counter-ions (g-ions), having the opposite sign to that of the fixed charge, is very high for reasons of electrical neutrality.

The result of these concentrations is that the membrane can easily be traversed by the g-ions, but is almost impermeable to the c-ions (see Figure 3.2).

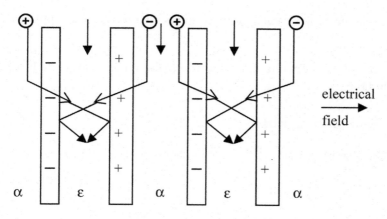

Figure 3.2. *Principle of electrodialysis*

An anionic membrane is negatively charged and allows cations to pass through. A cationic membrane allows anions to pass through. We also say that an anionic membrane is a cation "exchanger" membrane and that a cationic membrane is an anion "exchanger" membrane.

The feed comes in on all channels. The α channels are impoverished in terms of ions of both signs and the ε channels are enriched in ions of both signs. We then speak of the enriched solution (or the "concentrated" solution) and the impoverished solution (also known as the "diluted" solution).

That which we call "channels" are, in reality, very narrow (so as to decrease electrical resistance), and below, we shall employ the term "blades" instead. The thickness of the liquid blades is 0.5 to 2 mm and the velocity of the liquid is 0.5 m.s^{-1}, but may reach up to 2 m.s^{-1} if chicanes are introduced to make the path sinuous.

The pressure drop in a battery of hundreds of membranes may be of the order of 2–5 bars (5 bars if the path taken is sinuous).

Let us stress the fact that the liquid blades are arranged in parallel on the liquid side and serially on the side of the electrical current, as shown by Figure 3.2.

A so-called "elementary cell" contains two membranes and two liquid blades.

In general, the flowrate in an α blade is the same as in an ε blade, and is equal to Q.

3.9.2. Transport number

The transport number is the portion of the current that is transported by an ion:

$$t_{mi} = \frac{c_{mi} u_i z_i}{\sum\limits_{j-1}^{n} c_{mj} u_j z_j} \quad \text{when} \quad \sum\limits_{j} t_{mj} = 1$$

c_{mi}: concentration of the ion i in the membrane (kion.m^{-3})

u_{mi}: algebraic mobility of the ion i in the membrane (m^2.s^{-1}.V^{-1})

z_i: algebraic valence of the ion i

Numerous authors agree that for all the ions present in the membrane, the following ratio is the same:

$$\frac{u_{mi} \text{ in the membrane}}{u_{0i} \text{ in the external solution}}$$

Hence, the transport number is the same in the membrane and in the external solutions (at equal concentrations, of course).

Thus, we know (see section 3.2.1) that we have:

$$\frac{u_{mi}}{u_{0i}} = \frac{1}{t^2} \qquad \forall i$$

t: tortuosity of the pores: $1.4 < t < 1.8$

3.9.3. *Electrical resistance of the membrane*

The electrical current is due to the displacement of the ions under the influence of the electrical field. The flux density of the ion i in a pore is:

$$J_{0i} = -u_{0i}c_{0i}\frac{d\varphi}{dx}$$

[3.29]

u_{0i}: algebraic mobility of the ion i $(m^2.V^{-1}.s^{-1})$

c_i: concentration of the ion i $(kmol.m^{-3})$

φ: internal potential (V)

x: axial distance along the pore (m)

The index zero refers to the inside of the pore. If no index is present, then the concentration refers to the total volume of the membrane.

$$\frac{c_0}{c} = \frac{1}{\varepsilon_i}$$

ε_i: porosity accessible to the ion i

If ε is the geometric porosity of the membrane – i.e. the fraction of its volume occupied by the liquid – and because the center of an ion must be a distance greater than or equal to the radius of the ion away from the wall of the pores, we have:

$$\varepsilon_i = \varepsilon\left(1 - \frac{r_i}{r_p}\right)^2$$

r_i: radius of the ion (m)

r_p: radius of the pore (m)

The ionic flux density J is such that (see section 3.2.1):

$$\frac{J_{0i}}{J_i} = \frac{t}{\varepsilon_i}$$

t: tortuosity: $1.4 < t < 1.8$

Equation [3.29] is then written (with $L = tx$):

$$J_i \frac{t}{\varepsilon_i} = -u_i \frac{c_i}{\varepsilon} \frac{d\varphi}{tdx}$$

or indeed:

$$J_i = \frac{-u_{0i}}{t^2} c_i \frac{d\varphi}{dx}$$

Thus, the mobility u_i of an ion i in the membrane is equal to:

$$u_i = \frac{u_{0i}}{t^2}$$

The internal potential gradient from one face of the membrane to the other is:

$$\frac{d\varphi}{dx} = \frac{\varphi_2 - \varphi_1}{e_m} = \frac{-\Delta\varphi}{e_m}$$

e_m: thickness of the membrane (m)

The current density transported by the set of ions is:

$$i = \sum_i Fz_i J_i = \sum_i Fz_i u_i c_i \frac{\Delta\varphi}{e_m} = \frac{\Delta\varphi}{R_m}$$

$\Delta\varphi$: potential drop

i: $A.m^{-2}$

F: Faraday's constant: 96.487×10^6 $C.keq^{-1}$

z_i: algebraic valence of the ion

R_m: resistance of the membrane ($\Omega.m^2$)

A membrane whose surface area is Am^2 has a resistance equal to $(R_m/A) \ \Omega$.

Finally, the resistance due to all of the ions is:

$$R_m = \frac{e_m}{\sum_i z_i F u_i c_i}$$

or indeed:

$$R_m = \frac{t^2 e_m}{F \sum_j z_i u_{0i} c_i}$$

The displacement of the ions due to diffusion is very slight in comparison to the migration due to the electrical potential, so we can consider that the concentrations are invariable over the thickness of the membrane. In addition, these concentrations are obtained by the Donnan equilibrium on the basis of the upstream composition, so, for the carrying of the current, we can accept that only the g-ions are involved.

3.9.4. *Electrical resistance of a cell*

The electrical resistance of a cell is the sum of the resistances of the two membranes and those of the two liquid phases (enriched and impoverished).

The electrical resistance of a membrane is measured rather than calculated. More specifically, we measure the resistance corresponding to 1 m² of membrane, and that resistance is evaluated in $\Omega.m^2$. Of course, the resistance of a membrane whose surface area is A then is:

$$R = \frac{\rho_s}{A}$$

ρ_s: resistance of 1 m² of membrane ($\Omega.m^2$)

The resistance of the membrane is simply:

$$R_m = \frac{V}{I} = \frac{\rho_s}{A}$$

This gives us ρ_s.

In general, we work with concentrations corresponding respectively to the enriched solution and to the impoverished solution. The membrane resistance varies between $10^{-4}\ \Omega.m^2$ and $10^{-3}\ \Omega.m^2$. The thickness of the membranes lies between 0.1 and 0.5 mm.

The molar conductivity of an electrolyte present in a solution is (see section 4.6.4 in [DUR 16a]):

$$\Lambda_{mol} = \Sigma v_i |z_i| \lambda_{eqi} = \sum_i v_i \lambda_{ion\ i}$$

v_i: number of ions i in the formula of the electrolyte.

The conductivity of a solution containing c_j kmoles of electrolytes per m^3 is:

$$\kappa = \sum_j c_j \Lambda_{molj} \qquad (5.\ m^{-1})$$

The conductance of a liquid blade with surface area A and thickness e_L is:

$$G_L = \frac{\kappa A}{e_L}$$

Finally, the resistance of a cell containing two membranes, respectively negative and positive, and two impoverished and enriched liquid blades, respectively, is:

$$R_{cell} = \frac{1}{G_{La}} + \frac{1}{G_{Le}} + R_{m-} + R_{m+}$$

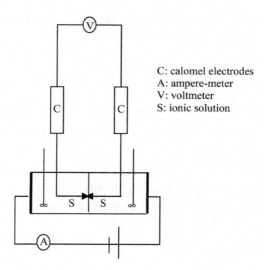

C: calomel electrodes
A: ampere-meter
V: voltmeter
S: ionic solution

Figure 3.3. *Measuring the resistance of a membrane*

The current passing through the cell is:

$$I = \frac{Q(|z_a|v_a + |z_c|v_c)F(c_f - c_s)}{}$$

Q: flowrate of dilute solution (impoverished) $(m^3.s^{-1})$

v_i: number of type-i ions stemming from the dissociation of the salt (stoichiometric coefficient)

c_f and c_s: concentrations of salt on entering and exiting the blade

ξ: current yield

NOTE.–

The variation in concentration of a solution over the height dH is:

$$dc_s = \frac{iL_m dH}{\Sigma z_i v_i FS}$$

L_m: width of the membrane (m)

The concentration varies in a linear fashion with the height H, so the mean value of $1/c$ in the expression of $1/G$ is:

$$\left(\frac{1}{c_d}\right)_{mean} = \frac{1}{c_s - c_f}\int_{c_s}^{c_f}\frac{dc_d}{c} = \frac{\ln\dfrac{c_f}{c_s}}{c_f - c_s}$$

3.9.5. Ion-transfer equation

The movement of the ions across the membrane due to any inequalities of concentration is negligible, and we can consider that only migration under the influence of the electrical field takes place.

Consider:

$$dA = L_m dH$$

dA: elementary surface area of a membrane

L_m: width of the membranes

Along a liquid blade whose height is dH, the variation dc_s in salt concentration is such that:

$$Qdc_s = J_s L_m dH$$

Q: flowrate of solution in the blade $(m^3.s^{-1})$

J_s: flux density of salt removed from a blade α or added to a blade ε $(kmol.m^{-2}.s^{-1})$

$$J_s = \frac{\xi i}{F\sum_i z_k v_k} = \frac{\xi i}{F\sum_j z_j v_j}$$

i: current density $(A.m^{-2})$

z: ionic valence

v: stoichiometric coefficient of the ion

k and j: indices for cations and for anions

Indeed, in a binary system, we know that:

$$(zv)_{cat} = (zv)_{ani} = zv$$

Hence:

$$dc_s = \frac{\xi i}{\sum zvFQ} L\, dH$$
[3.30]

The coefficient ξ is the current yield.

Whether we are dealing with type α (impoverished) or ε (enriched) blades, the thickness is constant and equal to e. Similarly, the liquid flowrate is the same and equal to Q.

Let us set:

c_α: local concentration of salt in the impoverished blade (kmol.m^{-3})

c_ε: local concentration of salt in the enriched blade (kmol.m^{-3})

The current density is:

$$i = \frac{U}{\dfrac{e}{\Lambda_s}\left(\dfrac{1}{c_\alpha} + \dfrac{1}{c_\varepsilon}\right) + R_{ma} + R_{mc}}$$
[3.31]

U: voltage drop in a cell (V)

Λ_s: molar conductivity of the salt (Ω^{-1}.kmol^{-1}.m^2)

R_{ma} and R_{mc}: resistances of the anionic- and cationic membranes

Let us eliminate i between equations [3.30] and [3.31].

$$dc_s = \cfrac{1}{\cfrac{e}{\Lambda_s}\left(\cfrac{1}{c_\alpha}+\cfrac{1}{c_s}\right)+R_{ma}+R_{mc}} \times \cfrac{LdH}{\Sigma z_j v_j FQ}$$

However, the concentration of the solute over the height of the liquid blades is written:

$$c_\alpha + c_\varepsilon = c_{\alpha f} + c_{\varepsilon f}$$

The index f characterizes the feed. Thus, we have:

$$c_\varepsilon - c_{\varepsilon f} = c_{\alpha f} - c_\alpha = \gamma$$

$$\frac{U\xi LdH}{\sum z_j u_j FQ} = \left[\frac{e(c_\alpha + c_\varepsilon)}{\Lambda_s c_\alpha c_\varepsilon} + R_a + R_c\right] dc_s$$

Let us multiply both sides of the equation by:

$$\frac{\Lambda_s}{e(c_\alpha + c_\varepsilon)} = \frac{\Lambda_s}{e(c_{\alpha f} + c_{\varepsilon f})}$$

and integrate from zero to H with:

$$dc_s = d\gamma$$

After simplification by $(c_{\alpha f} + c_{\varepsilon f})$, we obtain:

$$\left[-Ln(c_{\alpha f} - \gamma) + Ln(c_{\varepsilon f} + \gamma)\right]_0^\gamma + \frac{\Lambda_s(R_{ma} + R_{mc})\gamma}{e} = \frac{\Lambda_s U\xi LH}{\sum\limits_j z_j v_j FQ}$$

or indeed:

$$\ln\left(\frac{c_{\varepsilon s} \times c_{\alpha f}}{c_{\varepsilon f} \times c_{\alpha s}}\right) + \frac{\Lambda_s(R_{ma} + R_{mc})(c_{\alpha f} - c_{\varepsilon s})}{e} = \frac{\Lambda_s \xi UHL}{\sum\limits_j z_j v_j FQ}$$

This relation, supplemented by:

$$c_{\varepsilon s} = c_{\varepsilon f} + c_{\alpha f} - c_{\alpha s}$$

enables us to calculate $c_{\alpha s}$ on exiting the α blade, because the concentrations $c_{\alpha s}$ and $c_{\varepsilon s}$ are the concentrations at exit.

3.9.6. *Limiting current density*

In the membrane, the flux density of the g-ion is:

$$J_g = \frac{T_{mg} i}{F z_g}$$

i: current density (per unit surface of the membrane) (A.m^{-2})

F: Faraday: 96.487×10^6 C.kiloequivalent^{-1} (C.keq^{-1})

z_g: algebraic valence of the g-ion

T_{mg}: transport number of the g-ion in the membrane

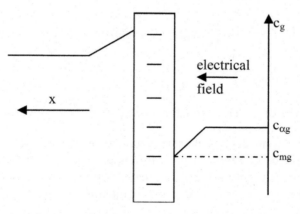

Figure 3.4. *Impoverished solution side of a negative membrane*

Consider the membrane on the side of the impoverished solution. The flux density of g-ions can be expressed:

– by the electrical current

$$\frac{T_{\alpha g} i}{F z_g}$$

$T_{\alpha g}$: transport number of the g-ion in the impoverished solution

– or else by the diffusion

$$-D_g \frac{dc_g}{dx}$$

We set:

$c_{\alpha g}$: concentration of the g-ion in the impoverished solution (kion.m^{-3})

c_{mg}: concentration of the g-ion at the surface of the membrane (kion.m^{-3})

D_g: diffusivity of the g-ion (m^2.s^{-1})

Let us state that the two values of J_g are equal, and integrate:

$$\left(T_{mg} - T_{\alpha g} \right) \frac{i}{F z_g} = -\frac{D}{\Delta x} \left(c_{mg} - c_{\alpha g} \right) = -\beta \left(c_{mg} - c_{\alpha g} \right)$$

T: transport number

β: coefficient of material transfer on contact with the membrane (m)

When the current density increases, c_{mg} must decrease and, at a limiting value of i, c_{mg} becomes zero. From that moment on, the current is transmitted only by the ions stemming from the dissociation of the water. This limiting value of the current density is obtained by making $c_{mg} = 0$.

$$i_{lim} = \frac{\beta F z_g c_{\alpha g}}{T_{mg} - T_{\alpha g}}$$

T_{mg} is always greater than $T_{\alpha g}$, because the concentration of the g-ion in the membrane is much greater than it is in the impoverished solution.

Beyond the limiting current density, the differential conductance di/dV of a cell is constant but low. However, when the potential difference is high (overlimiting current density), the conductance di/dV again takes a constant, and even lower, value (see [KRO 98]).

3.9.7. Transference and permselectivity of a membrane

The transference of the ion i is, by definition, the ratio:

$$t_{mi} = \frac{c_{mi}\left|u_i\right|}{\displaystyle\sum_{j=1}^{n} c_{mj} u_j z_j}$$

For an electrolyte containing two monovalent ions:

$$t_{m+} + t_{m-} = 1 = T_{m+} + T_{m-}$$

By definition, the permselectivity of the g-ion is:

$$\psi_{mi} = \frac{t_{mi} - t_{0g}}{1 - t_{0g}}$$

t_{mg}: transference of the g-ion i in the membrane

t_{0g}: transference of the g-ion i in the feed solution

For example, a cation exchanger membrane – i.e. a negatively charged membrane – will have a permselectivity equal to 1 if the transference T_{mg} is equal to 1 for the cation (which is the g-ion). This means that, in the membrane, the concentration of the other ions is close to zero.

If the concentrations in the membrane are equal to those of the feed solution, the permselectivity will be zero.

The permselectivity varies, from case to case, between 0.7 and 0.95.

NOTE.–

If a solution contains only one solute and two types of ions, a and c:

$$T_{0a} + T_{0c} = 1$$

and, for the cation, for example, the permselectivity is:

$$\psi_{mc} = \frac{T_{mc} - T_{0c}}{T_{0a}}$$

3.9.8. *Transport number of water*

One might think that the difference in concentrations between the dilute (impoverished) solution and the concentrated (enriched) solution would, by osmosis, cause water to cross the membrane from the impoverished solution toward the enriched solution. In fact, this phenomenon is not particularly significant.

In reality, the two causes of water transfer are:

– the entrainment of water by ions which are surrounded by water molecules – i.e. hydrated;

– the movement of the ions within the membrane causes a mechanical entrainment of the water in the vicinity of these ions, which is the electro-osmotic flux.

Finally, the flux density of the water molecules is:

$$J_e = T_{me} \sum_i J_i$$

T_{me}: "transport number" of water (kmol of water. kion^{-1})

J_i: flux density of the ion i crossing the membrane (kion.m^{-2}.s^{-1})

In general, the ions i are attached to a single g-ion.

The transport number of the water is of the order of 4–8.

The flux density devoted to the transfer of water in a cell containing two ion-exchange membranes – one an anion-exchange membrane and the other a cation-exchange membrane – is:

$$J_e = (T_{mae} + T_{mce})\frac{Q}{A}(c_f - c_s) \qquad (kmol.m^{-2}.s^{-1})$$

Q: flowrate of solution (impoverished or enriched, $(m^3.s^{-1})$)

A: surface area of a membrane (m^2)

The current density is, theoretically:

$$i = \frac{kFQ}{A}(c_f - c_s) \qquad \left(C.m^{-2}.S^{-1}\right)$$

Thus:

$$\frac{J_e}{i} = \frac{(T_{mae} + T_{mce})}{F}$$

This ratio is approximately 10^{-9} kmol.C^{-1}.

3.9.9. *Current yield*

The ideal current density is:

$$i_{id} = z_g F(c_{0g} - c_{\alpha g})\frac{Q}{A} \qquad (A.m^{-2})$$

Q: flowrate of the feed solution $(m^3.s^{-1})$

c_{0g} and $c_{\alpha g}$: concentrations of g-ions respectively in the feed solution and in the impoverished solution $(kmol.m^{-3})$

z_g: valence of the g-ion

F: Faraday: 96.487×10^6 C.keq^{-1}

The real current density is:

$$i_{real} = \frac{i_{id}}{\zeta}$$

ξ: current yield

The current passing through an electrodialytic battery is not used entirely for the transport of the g-ions.

– the first portion of the current serves for the transport of the c-ions toward the impoverished phase, particularly when the concentration of the feed reaches a value of the same order as that of the fixed charges of the membrane. The corresponding yield is:

$$\eta_c = 1 - T_{mc} = T_{mg}$$

T_{mc}: transport number of the c-ions in the membrane

– a second part pertains to reverse transport due to osmosis – i.e. to the difference in concentrations on both sides of the membrane. This osmosis transports the g-ions in the opposite direction to the electrical field, and thus partly cancels out the effect of the electrical current. This corresponds to the yield η_0.

The yields η_c and η_0 are not too far from 1.

On the crossing of the manifold, current losses are experienced which are characterized by a yield η_m which may be as high as 0.95.

The Faraday yield is defined by:

$$\eta_F = \eta_m \left[(T_{mc} - T_{0c}) + (T_{ma} - T_{0a}) \right] \eta_c \eta_0$$

Let us introduce the permselectivities (see section 3.9.7).

$$\eta_F = \eta_m \left(\psi_{mc} T_{0a} + \psi_{ma} T_{0c} \right) \eta_c \eta_0$$

If, as an initial approximation, we accept that:

$$T_{0a} \# T_{0c} = 0.5$$

We obtain an expression of the Faraday yield:

$$\eta_F = \eta_m \frac{(\psi_{mc} + \psi_{ma})}{2} \eta_c \eta_0$$

The entrainment of water by the ions slows them down and, for a given potential drop, the intensity obtained is lower, giving us the yield:

$$\eta_e = \left[1 - k(T_{mae} + T_{mce}) \right]$$

k: empirical coefficient

η_e: yield because of water transfer

Finally, the current yield of the cell is:

$$\xi = \eta_F \eta_e = \eta_m \eta_c \eta_0 \frac{(\psi_{mc} + \psi_{ma})}{2} \left[1 - k(T_{mae} - T_{mce}) \right]$$

where:

$$v_g z_g F(c_f - c_s) Q_d = \xi I$$

v_g and z_g are the stoichiometric coefficient and the valence of the g-ion

Q_d: flowrate of impoverished (dilute) solution in a cell ($m^3.s^{-1}$)

c_f and c_s: salt concentration of the solution to be impoverished at the inlet and outlet of a cell ($kmol.m^{-3}$)

In relation to the surface of a membrane, the current density is of the order of 2000 $A.m^{-2}$.

3.9.10. *Bipolar membranes*

A bipolar membrane is composed of two adjacent membranes, separated by a liquid layer whose thickness is around 2–4 nm. In that thin layer, the water dissociates, giving rise to OH⁻ and H⁺ ions.

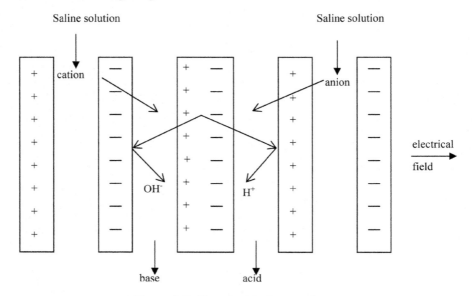

Figure 3.5. *Use of a bipolar membrane*

From the energy point of view, the bipolar membrane is an economical means of electrolyzing a salt.

The Gibbs energy expended by the action of the bipolar membrane is:

$$\Delta G = RT \times 2.3 \left(pH_{base} - pH_{acid} \right) = RT \times 2.3 \, \Delta pH$$

Fractionation of Dispersions by Natural Decantation (by Gravity)

4.1. Decantation rate

4.1.1. *General*

When an isolated particle submerged in a fluid is subjected to a force field (gravitation, electrostatic force, etc.), it moves.

The force due to the field is soon counteracted by the friction exerted by the fluid on the particle, and the particle is then animated with a velocity known as the limiting velocity.

When the particle is not alone, it and its neighbors all move at the same speed, which is slower than that of a lone particle.

In all problems of decantation in a liquid, the important factor is the difference between the weight and the Archimedes upthrust. Thus, we use the difference between the densities of the solid and of the liquid, which is justified even if the force field at work is not gravity, but a different force. We then speak of the "effective density".

Before going any further, let us specify that the function of a decanter is to separate a pulp (mash) into a sediment (the mud) and a clear liquor.

– *clarifiers* are designed to evacuate the solid dispersed in the liquid to obtain a clear liquor;

– *thickeners* are designed to obtain a sediment (the mud) which is as compact as possible – i.e. which has as high a content of solid material as possible.

4.1.2. *Limiting velocity of a lone particle*

Let g represent the acceleration resulting from the influence of the force field.

We propose to directly determine the limiting velocity of the particle in the following manner:

$$\text{calculate } X = \frac{4g\rho_F d_p^3 \left(\rho_p - \rho_F\right)}{3\mu_F^2}$$

From this, we can deduce Y by one of the following relations:

$$1 < X < 10^4 \qquad\qquad Y = \frac{13.842}{X^{1.97}} + \frac{218}{X^{1.074}}$$

$$10^4 < X < 10^9 \qquad\qquad Y = \frac{59.11}{X^{0.923}} + \frac{0.0278}{X^{0.378}}$$

The limiting velocity then is:

$$V_\ell = \left[\frac{4g\mu_F \left(\rho_p - \rho_F\right)}{3\rho_F^2 Y}\right]^{1/3}$$

ρ_p: density of the particle (kg.m^{-3})

ρ_F: density of the fluid (kg.m^{-3})

d_p: diameter of the particle, supposed to be spherical (m)

μ_F: dynamic viscosity of the fluid (Pa.s)

g: acceleration (m.s^{-2})

V_ℓ: limiting velocity (m.s^{-1})

EXAMPLE.−

$$d_p = 30 \times 10^{-6} \text{ m} \qquad\qquad \rho_p = 2700 \text{ kg.m}^{-3}$$

$$\mu_F = 10^{-3} \text{ Pa.s} \qquad\qquad \rho_F = 1000 \text{ kg.m}^{-3}$$

$$X = \frac{4 \times 9.81 \times 1000 \times \left(30 \times 10^{-6}\right)^3 \times 1700}{3 \times 10^{-6}} = 0.6$$

$$Y = \frac{13842}{0.366} + \frac{218}{0.578} = 37819 + 377$$

$$Y = 38196$$

$$V_\ell = \left[\frac{4 \times 9.81 \times 10^{-3} \times 1700}{3 \times 10^6 \times 38196} \right]^{1/3} = 0.000835 \text{ m.s}^{-1}$$

4.1.3. *Division of the height between the clear liquid and the sediment*

Suppose we are dealing with the decantation of a given load, as is the case in a test chamber.

When the height of clear liquor has risen by Δh_c, the height of sediment, for its part, has increased by Δh_s.

At the top, the liquid released by the solid is:

$$\Delta h_c \left(1 - \varepsilon_B\right)$$

At the bottom, the liquid expelled by the solid is:

$$\Delta h_s \left(\varepsilon_B - \varepsilon_s\right)$$

ε_B: porosity of the original mash to be decanted

ε_s: porosity of the sediment

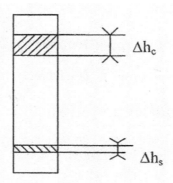

Figure 4.1. *Decantation of a mash in a test chamber*

These two quantities of liquid are equal, because over the initial height H of the mash, the total quantity of liquid has not varied. Thus, at the end of the operation, we have:

$$h_c + h_s = H \quad \text{and} \quad \frac{h_c}{h_s} = \frac{\varepsilon_B - \varepsilon_s}{1 - \varepsilon_B}$$

From these two equations, we draw:

$$h_c = H\left(\frac{\varepsilon_B - \varepsilon_s}{1 - \varepsilon_s}\right) \quad \text{and} \quad h_s = H\left(\frac{1 - \varepsilon_B}{1 - \varepsilon_s}\right)$$

EXAMPLE.–

$$\varepsilon_B = 0.95 \qquad\qquad \varepsilon_s = 0.4$$

$$\frac{h_c}{H} = \frac{0.95 - 0.4}{1 - 0.4} = 0.917 \qquad\qquad \frac{h_s}{H} = \frac{1 - 0.95}{1 - 0.4} = 0.083$$

NOTE.–

The decantation rate is defined by:

$$V_{SA} = \frac{dh_c}{d\tau}$$

V_{SA} is the absolute velocity of the solid, which is its velocity in relation to the space.

4.1.4. *Friction velocity of liquid-on-solid*

This velocity is a relative velocity:

$$V_{RF} = V_{SA} + \frac{U_{FT}}{\varepsilon_B}$$

However, we know that the velocity in an empty bed U_{FT} of the liquid coming back up is:

$$U_{FT} = \frac{\Delta h_c}{\Delta t}(1 - \varepsilon_B) = V_{SA}(1 - \varepsilon_B)$$

and the (relative) velocity of friction of the liquid on the solid becomes:

$$V_{RF} = V_{SA}\left[1 + \frac{1 - \varepsilon_B}{\varepsilon_B}\right] = \frac{V_{SA}}{\varepsilon_B}$$

NOTE.–

In the case of fluidization, overall, the solid is immobile in relation to the chamber, and we simply have:

$$V_{RF} = \frac{U_{FT}}{\varepsilon_B}$$

NOTE.–

Let us accept that for a given value of V_{RF}, there is a corresponding exact value of ε_B, and vice versa.

We then see Richardson and Zaki's conclusion [RIC 54], which was as follows (for an equal value of V_{RF}):

$$V_{SA} = U_{FT}$$

The falling velocity of a suspension with given porosity in relation to a fixed horizontal plane is equal to the velocity in an empty bed of the liquid, which keeps the suspension fluidized at the same porosity.

4.1.5. Decantation velocity of a particle cloud

For liquids, this rate V_{SA}, which is assessed in relation to a fixed horizontal plane, can be calculated by Richardson and Zaki's correlation [RIC 54]. This correlation pertains to a population of particles, all of which are of the same size.

$$V_{SA} = V_{\ell}\varepsilon^n$$

We calculate the Reynolds number:

$$Re = \frac{V_{\ell}d_p\rho_L}{\mu_L}$$

ρ_L and μ_L: density (kg.m^{-3}) and viscosity (Pa.s) of the liquid

d_p: diameter of the particles (m)

V_{ℓ}: limiting freefall velocity of a lone particle (m.s^{-1})

where:

$Re < 1$	$n = \dfrac{4.35}{Re^{0.03}}$
$1 < Re < 500$	$n = \dfrac{4.45}{Re^{0.03}}$
$Re > 500$	$n = 2.39$

EXAMPLE.–

Decantation in water:

$V_\ell = 0.000837 \text{ m.s}^{-1}$ $\qquad$ $\mu_L = 10^{-3} \text{ Pa.s}$ $\qquad$ $d_p = 30 \times 10^{-6} \text{ m}$

$\rho_L = 1000 \text{ kg.m}^{-3}$ $\qquad$ $\varepsilon = 0.95$

$$\text{Re} = \frac{0.00087 \times 30 \times 10^{-6} \times 1000}{10^{-3}} = 0.025$$

$$n = \frac{4.45}{0.025^{0.03}} = 4.97$$

$$V_{SA} = 0.95^{4.97} \times 0.000837 = 0.000648 \text{ m.s}^{-1}$$

4.2. Decanters in the chemical industry

4.2.1. Types of decanters

Decanters may be:

– either clarifiers, if they are used essentially to clarify the liquid phase;

– or thickeners if they are used to increase the compacity $\beta = 1 - \varepsilon$ of the mash in question, where ε is the volumetric fraction of liquid.

From the standpoint of the device's operation, we can distinguish:

– batch devices, in which the load is left at rest until decantation is completed. This mode of operation is rare;

– continuous-flow piston decanters, in the shape of a parallelepiped, in which the clear liquor, the dispersion and the sediment are supposed, for the purposes of the calculation, to move at the same horizontal velocity along the longer side. Parallelepipedic sand-removers used to rid water of the sand that it contains are similar to piston decanters. The calculations for the piston decanter are run in the same way as for the batch decanter whose length of stay would be the quotient of the length by the advancement rate;

– countercurrent continuous-flow decanters where the liquid rises against the flow of the descending solid. Flat-bottomed circular devices, fed in the center, are similar to this type of device. The overflow (clear liquor) collected at the periphery and the underflow (sediment) is collected in the center (Figure 4.4).

4.2.2. *Test in a test chamber*

Prior to designing any industrial decanters, it is essential to carry out a decantation experiment in a test chamber.

A dispersion of polydispersed particles is left at rest in a graduated test chamber. After a certain period of time, we observe three zones in the test chamber:

– the clear liquor at the invariable level h_o;

– the dispersion (mash) at the level h_B;

– the sediment at the level h_S.

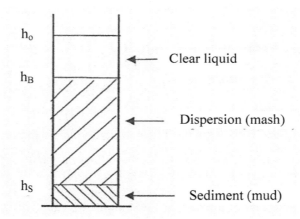

Figure 4.2. *Decantation in a test chamber*

4.2.3. *Characteristic curves of decantation (extrapolation)*

Characteristic curves represent the variation, over time, of the height of the interface separating the mash and the clear liquor.

To begin with, the decantation rate $\dfrac{dh_B}{d\tau}$ is constant. In other words, the decantation curve is a straight line, until the mash has completely disappeared. At that moment, the clear liquor is in direct contact with the sediment. From that moment on, the phase of piling of the sediment takes over.

We shall describe the characteristic of decantation using Kynch's hypothesis ([KYN 52], see section 4.2.7): the vertical motion of a horizontal layer depends only on the concentration (or, which is tantamount to the same thing, on the volumetric fraction of solid) in that layer. In reality, the velocity dh/dτ also depends on the fall velocity of a lone particle (see section 4.1.5) – at least when we are operating within the slurry itself.

In the sediment, matters are more complex, and can involve the elasticity of the sediment under compression, but this does not alter the validity of Kynch's hypothesis. As the sediment piles up, the area where the concentration is constant and equal to c rises from the bottom, at a rate U. The solid being deposited in that zone comes from a layer at the concentration c − dc with a decantation rate V + dV in relation to the recipient, but with a velocity V + dV + U in relation to the layer of concentration c. The solid leaves that layer (at the concentration c) with the velocity V + U.

The concentration c being constant, what exits and enters that layer is equal and, per unit time and horizontal surface area, we find:

$$(c-dc)(V+dV+U)=c(V+U)$$

Let us overlook the second-order infinitely small terms. We obtain:

$$U=c\frac{dV}{dc}-V \quad \text{where} \quad V=f(c) \quad \text{and} \quad \frac{dV}{dc}=f'(c)$$

Finally, the rise rate U of the layer of concentration c is constant, as indeed is that concentration. Note that V is orientated downwards and U upwards – hence the minus sign.

In Figure 4.3, the "curve" AB is a straight line, and the triangle AOB is the zone where the pulp exists. On the arc BD, the sediment piles up, and at D, it has reached its final thickness. The lines OB, OC and OD illustrate the rise of the layers of concentrations c_B, c_C and c_D.

After the final period τ_f, the sediment has reached its ultimate thickness h_f. In reality, the sediment would continue to increase slightly in bulk if we were to let it deposit beyond the time τ_f, but to do so is unacceptable in economic terms, and thus the time τ_f is an economical optimum.

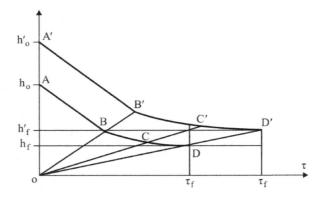

Figure 4.3. *Characteristic curves of decantation*

Figure 4.3 shows a second characteristic curve A' B' C' D', starting from a height h'_o which is greater than h_o. The points OBB' are aligned, because they correspond to the same concentration and thus to the same velocity $U_B = U_{B'}$. The same is true for the points OCC' and ODD'. The tangents to the curve BCD at the points B, C and D have the same slope as the tangents to the curve B' C' D' at the points B', C' and D', as these slopes measure equal decantation rates V. Indeed, the concentrations are equal at B and B', at C and C' and at D and D'. Finally, the two characteristic curves can be deduced from one another by homothety with the homothetic center 0.

The homothetic ratio is:

$$\frac{\tau'_f}{\tau_f} = \frac{h'_o}{h_o}$$

NOTE.–

The linear evolution AB corresponds to the *clarification* of the liquor and the evolution BCD shows the settling of the sediment – i.e. the *thickening* of the mud.

The theory behind thickening is given in section 51.1.2.

NOTE.–

According to Coe and Clevenger [COE 16], the situation in practice may differ somewhat from the above theory. For this reason, those authors recommend performing the tests in a tube 10 cm in diameter and 3 m tall.

4.2.4. Horizontal surface of a decanter

The instantaneous decantation rate $\overline{V}$ is:

$$\overline{V} = \left|\frac{dh}{d\tau}\right| > 0$$

The mean value of that velocity over the period τ_f is:

$$\overline{V} = \frac{1}{\tau_f} \int_o^{\tau_f} \left|\frac{dh}{d\tau}\right| d\tau = \frac{h_o - h_f}{\tau_f}$$

In Figure 4.3, we see that $\overline{V}$ is the slope of the line AD.

In view of the homothety of the characteristic curves of decantation, the value of $\overline{V}$ does not depend on the height of the test tube. However, a skeptical user could, firstly, perform a test with a test tube 50 cm tall and, secondly, as Coe and Clevenger [COE 16] did, perform a test with a tube 3 m tall.

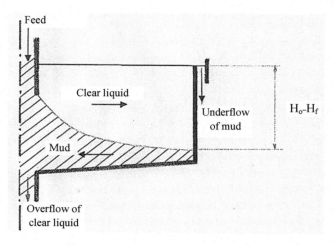

Figure 4.4. *Half-section of a decanter*

The volume of liquid Ω_L released per kg of dry solid is:

$$\Omega_L = (X_A - X_D)/\rho_L$$

X_A and X_D are the "humidities" on dry basis of the feed and of the mud in the underflow.

$$X = \frac{\text{mass of liquid}}{\text{mass of solid}}$$

If W_s is the flowrate of solid to be treated in terms of mass (e.g. the flowrate of ore), the flowrate of clear liquid to be produced is:

$$Q_L = W_s \Omega_L$$

That liquid flow must cross over the horizontal surface of the decanter at the velocity $\overline{V}$. From this, we deduce that surface Σ:

$$\Sigma = \frac{Q_L}{V} = \frac{W_s \Omega_L}{V} = \frac{W_s}{\overline{V}\rho_L}(X_A - X_D)$$

Figure 4.4 shows the vertical half-section of a decanter.

4.2.5. Recap

Considering a decantation curve, it is possible to give more simply an expression of the above results pertaining to the calculation of a decanter.

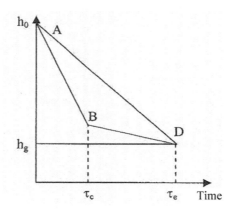

Figure 4.5. *Decantation in a thickener*

In Figure 4.5, which is a classic decantation curve, the point A corresponds to the initial state and the point D to the final state. The mean velocity which we have just examined is a fictitious velocity which is the slope of the line AD. Thus, the calculations for a thickener and a clarifier are performed in the same way, choosing:

– for the clarifier, the slope of the line AB, because a simple clarification does not include thickening of the sediment (curve BD);

– for the thickener, the slope m of the line AD. The point D corresponds to a given thickening of the mud.

$$\Sigma = \frac{Q_L}{m} \qquad [4.1]$$

Q_L: flowrate of clear liquid to be obtained

Let us give a simple justification for that relation. The volume of the decanter is:

$$h_0 \Sigma = Q_F \tau \qquad [4.2]$$

Q_F: feed flowrate $(m^3.s^{-1})$

τ: residence time of the feed (s)

Σ: surface area of the decanter

The volume of clear liquid put out is:

$$Q_L = \frac{h_0 - h_f}{h_0} Q_F \qquad [4.3]$$

Term by term, let us divide equation [4.2] by equation [4.3].

$$\frac{h_0 \Sigma}{Q_L} = \frac{\tau h_0}{h_0 - h_f} = \frac{h_0}{m}$$

This gives us equation [4.1].

4.2.6. *Height of a decanter*

The height of the decanter is the sum of three components:

– the feed enters at a depth of 50 cm from the free surface to prevent pollution of the clear liquid, which exits through the overflow;

– a height of 50 cm is left at the bottom of the decanter for the rakes which bring the mud into the center of the device;

– a height of 1–3 m is left for the zone where decantation takes place – i.e. for the active zone. Indeed, the clear liquid and the mud are endowed with opposite radial velocities, and the resulting velocity gradient must be reduced to the minimum to prevent untimely mixing between the clear liquid and the mud. The maximum of this velocity gradient is situated in the vicinity of the axis of symmetry of the circular thickener (see Figure 4.4).

4.2.7. *Flux density of the solid [KYN 52]*

Let β represent the compacity of a dispersion or of a sediment. The compacity is the ones' complement of the porosity; it is the volumetric fraction of the solid phase.

According to Richardson and Zaki [RIC 54], the sedimentation rate is (see section 4.1.5):

$$V = V_\ell (1-\beta)^n \qquad\qquad (\text{where } n > 2.39)$$

To obtain the volumetric flux density in accordance with Kynch [KYN 52], we merely need to multiply V by β.

$$\varnothing_s = V_\ell \beta (1-\beta)^n$$

For a determinate system (viscosity, densities, diameter of the particles, and therefore limiting velocity V_ℓ), we note that $\varnothing_s$ depends only on the compacity β (equal to $1 - \varepsilon$, where ε is the porosity) – i.e. on the volumetric fraction in solid. Here we see the hypothesis that Kynch had inadvertently formulated earlier on.

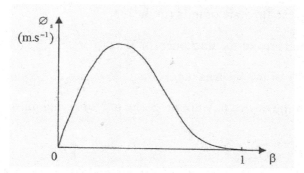

Figure 4.6. *Solid flux density as a function of the compacity*

Kynch indulges in considerations on discontinuities of concentration (of compacity) along a vertical. We shall not linger on this here.

In practice, the flux density in terms of mass (no longer in terms of volume) is obtained by multiplication by the density ρ_s of the solid.

$$G_{dec} = \rho_s V_\ell \beta (1-\beta)^n \qquad \qquad \left(kg.m^{-2}.s^{-1}\right)$$

For ores, in general, the *productivity* of the thickener is such that:

$$500 \, kg.m^{-2}.day^{-1} < G_{dec} < 1700 \, kg.m^{-2}.day^{-1}$$

β is of the order of 0.4 or 0.5 in the mud removed from the thickener.

Note that, for the biological muds used for water treatment, the true density is close to that of water and V_ℓ is very low, so we have:

$$G_{dec} \# 50 \, kg.m^{-2}.day^{-1}$$

Let us define the flux density of mud removal by:

$$G_{remov} = \frac{\rho_s Q_B \beta}{A}$$

ρ_s: density of the solid $(kg.m^{-3})$

Q_B: volumetric flowrate of mud $(m^3.h^{-1})$

A: horizontal area of the thickener (m^2)

β: compacity of the suspension

In order for the device to work properly, the following must be true:

$$G_{remov} = G_{dec}$$

If $G_{remov} > G_{dec}$, then the compacity of the mud will be decreased.

If $G_{remov} < G_{dec}$, then we will see an increase in the level of mud in device, the compacity of the removed mud will be increased and the clear liquid may be disturbed.

NOTE.–

Certain authors agree that the slope of the decantation curve does not exhibit a discontinuity at the interface separating the pulp from the sediment. With this assumption made, those same authors are able to indulge in hypothetical discussions involving fictitious concentrations. Here, this outlook has been abandoned (see [TAL 55]).

4.2.8. *Torque of the scraper (circular decanters)*

The role of the scraper is twofold:

– to facilitate the evacuation of the sediment;

– to facilitate the release of the interstitial liquid and possibly of occluded gases, through a vertical harrow attached to the scraper.

The peripheral velocity V_P of the scraper is of the order of 10 m.mn^{-1}.

The rotation speed given by the reducer is:

$$N = \frac{V_p}{2\pi R_D} \qquad \left(rev.mn^{-1}\right)$$

Here, R_D is the radius of the decanter, rather than that of the scraper.

The mechanical control of the scraper must be robust, and the reducer must be able to resist a significant amount of torque. Let us examine this point.

The elementary torque corresponding to a length of arm dR situated at the distance R from the axis is proportional to three values:

1) The velocity, because we are in the laminar regime. If V_p is the peripheral velocity (counted as though the arm had the same radius as the decanter), the velocity at the distance R is:

$$V_p \frac{R}{R_D}$$

2) The distance from the axis R.

3) The surface of the scraper element h dR, where h is the equivalent height of the arm such that h dR represents the projection of the scraper element on a vertical plane perpendicular to its motion.

Thus:

$$dC = \frac{KV_p R^2 h dR}{R_D}$$

By integration, we find that the unitary torque for an arm is:

$$C_u = \frac{KV_p h R_R^3}{3R_D}$$

where R_R is the length of the arm of the rake.

and for n arms (generally 2 or 4):

$$C = \frac{nKV_p h R_R^3}{3R_D} \quad \text{where} \quad K = 18,000 \text{ kg.m}^{-2}.s^{-1}$$

(This value of K is merely an order of magnitude).

The power of the motor is given by:

$$P = \frac{C.V_p}{1000\ R_D}$$

C: torque (N.m)

V_p: peripheral velocity (m.s^{-1})

R_D: radius of the decanter (m)

P: power (kW)

EXAMPLE.–

$V_p = 10$ m.mn^{-1} = 0.17 m.s^{-1} $R_D = R_R = 20$ m

$n = 2$ $h = 0.7$ m

$$C = \frac{2 \times 18,000 \times 0.17 \times 0.7 \times 20^3}{3 \times 20}$$

$$C = 571,200\ \text{N.m}$$

$$P = \frac{571,200 \times 0.17}{1000 \times 20}$$

$$P = 4.9\ \text{kW}$$

4.3. Decanters in ore extraction

4.3.1. Grooved tables

We pour the pulp and wash water over the table, which is slightly inclined from the horizontal. The heavy particles become deposited in the grooves, which form an angle of a few degrees with the cross-pieces at the level of the table. The lightweight particles jump over the partitions of the grooves.

So as to continuously recover the heavier material, we vibrate the table in a direction perpendicular to the flow, but the cycle is asymmetrical. One change of direction is progressive, whereas the other is sharp, so that the heavy particles progress along the grooves in the direction of the triangle A, which thus recovers the content of the grooves. The frequency of the vibrations is of the order of 5–6Hz and their amplitude varies from 1 to 3 cm.

The triangle B corresponds to the feed with pulp and wash water:

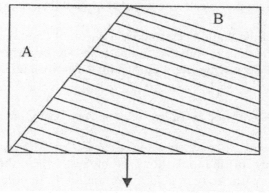

Flow of pulp and wash water

Figure 4.7. *Grooved table*

Note that it would perhaps be more appropriate to speak of wrinkled tables than grooved ones, because the grooves are fine and are close together.

4.3.2. *Multi-component hydraulic sorter*

These devices present the greatest efficiency in terms of separation.

At the base of each compartment (there may be up to 8 compartments), we inject water at a low pressure, which causes the gradual sedimentation of the particles. The larger particles are recovered at the bottom, and the finer ones spill over into the next compartment.

In each compartment, the operator must carefully choose the flowrate of water.

4.3.3. *Jigs*

In this context, the word "jig" refers to a calibration device.

Jigs are used to separate particles of different densities using a pulsed liquid stream. The pulsations take place in the vertical direction, which causes stratification:

– the heavy particles move toward the bottom;

– the lightweight particles come together around the top.

The motion of the liquid is created by a piston of large diameter, situated on one side of the device. Initially, the particles are deposited on a sheet of perforated metal occupying the majority of the horizontal cross-section of the device.

These devices are used to isolate ores or very heavy metals from sterile materials.

The frequency of the pulsations is highest when the particles are sedimentary-type particles (e.g. gold, tin or tungsten). In order for the device to work properly, the ore to be recovered needs to be very heavy (e.g. galenite, cassiterite, barite, etc.).

It is possible to create jigs where it is the perforated plate which vibrates rather than the liquid, but the mechanical making of such devices is apparently more complicated.

4.3.4. *Classification by dense liquor (gravimetric sorting)*

It is not often that we find a liquid which has the right density to separate the particles which *sink* and the particles which *float*. It is possible to overcome this difficulty by using a dispersion, in water, of fine particles which, as they are animated with Brownian motion, do not fall out of suspension as sediment. If ϕ is volumetric fraction of solid, the density of the dispersion is:

$$\rho_D = \rho_E (1 - \phi) + \rho_s \phi$$

This process is appropriate if:

$$\left| \rho_{S1} - \rho_{S2} \right| \geq 100 \ kg.m^{-3}$$

The dispersion is made with magnetite and/or ferro-silicon (or galenite).

The dispersed product must be:

– dense: magnetite (d = 5.15), galenite (d = 7.5), ferro-silicon (d = 7),

– of limited viscosity. Thus, the volumetric fraction must be such that:

$$\phi \leq 0.35$$

– hard: the grains must not break, as this could hamper their recovery;

– stable in dispersion. Put differently, the decantation rate V must be such that:

$$V < 1.7 \times 10^{-5} \ m.s^{-1}$$

The magnetite and ferro-silicon are recovered by a process using magnetism. A sulfide such as galenite is recovered by flotation (see Chapter 6). In certain cases, it is possible to use the ore itself as a dispersed solid, in which case we do not need to recover it.

The simplest device is a downward-pointing cone. Both the ores and the dense liquor are fed in at the top. The sinking fraction is recovered at the bottom by a membrane pump, and the floating fraction is removed through the overflow.

In certain devices, the sinking fraction is entrained at the bottom by a bladed transporter, which brings it out of the decanter and leaves it at a higher level than that of the free surface of the dense liquor.

4.3.5. Helical differential decantation (Humphrey system)

The trough generally makes five rotations with a step of 35 cm.

The optimum size of the heavy particles must be between 2 and 0.07 mm.

X lightweight particles (sterile)

● heavy particles (ore)

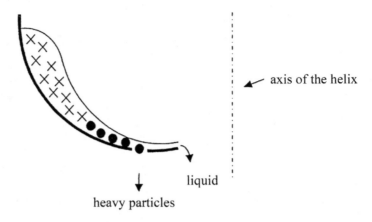

Figure 4.8. *Helical trough section*

The heaviest particles are evacuated through 15 orifices throughout the device, and situated at the bottom of the cross-section of the trough. These orifices can be entirely or partially blocked by special bungs. They are distributed every 120°.

The liquid and the light particles completely follow the spiral path, but the liquid spills over partially in the direction of the axis.

Small orifices are situated between the orifice of evacuation of the heavy materials and the liquid outlet. They serve to admit wash water to rid the heavy particles of any light ones which they may have entrained.

Solids that are sufficiently heavy to be dealt with by this device include, in particular: chromite, ilmenite, zirconia, rutile, tin, iron ore and certain phosphates.

The spiral can absorb 1.5 tonne of feed per hour, and up to 100 spirals can operate in parallel.

Destabilization of Sols (Colloidal Dispersions), Coagulation and Flocculation

5.1. Colloidal dispersions: the double layer

5.1.1. *Introduction*

Depending on the size of the particles involved, we distinguish between:

– true solutions and macromolecular "solutions" where the size of the molecules of solute d_s is less than 10 nm;

– colloidal and microfine suspensions where the size of the particles is situated between 10 nm and 1 µm. Thus, particles of iron hydroxides are generally smaller than 1 µm. These suspensions do not decant, so we need to try to treat them by coagulation or flocculation. Indeed, simple Brownian motion is sufficient to prevent any sedimentation. The term "sol" describes an intermediary system between a true solution (limpid) and a dispersion which diffuses light to a greater or lesser degree, and is therefore more or less opaque. The size of the colloids is of the order of magnitude of the wavelengths of visible light, which accounts for the diffusion of light by sols.

In general, colloids carry an electrical charge at their surface, and that charge is the same for all these particles, which therefore repel one another. Hence, there can be no clumping in the natural state. However, by adjusting the pH or by adding an ionic adjuvant, or a polymer, we shall see that it is possible to cause colloids to clump.

5.1.2. *Charge, surface potential and ζ potential*

When dispersed in water, colloids generally have an electrical charge on their surface. That charge may be due, for example:

– to ions leaving the crystalline lattice, leaving a charged vacancy (silver iodide presents an example);

– to reactions of the surface groups with the surrounding water. Thus, metal oxides, depending on the pH, produce the following reactions:

$$M-OH+Na^+ +OH^- \rightarrow M-O^- +Na^+ +H_2O \quad \text{basic pH} > 7$$

$$M-OH+H^+ +NO_3^- \rightarrow M-OHH^+ +NO_3^- \qquad \text{acidic pH} < 7$$

The ions thus fixed to the surface, with their corresponding charge, are known as the *determinant* ions. They are held by so-called *generic* bonds.

Gouy [GOU 10, GOU 17] and Chapman [CHA 13] showed that the charges thus appearing were compensated within the liquid by charges of the opposite sign which form a so-called diffuse layer.

$$\sigma_o + \sigma_d = 0$$

σ_o: surface charge density of the determinant ions $(C.m^{-2})$

σ_d: diffuse charge density (in relation to the surface area of the colloid) $(C.m^{-2})$

The charges in the diffuse layer constitute the "counter-charges". However, a portion of the counter-charges may be bound to the surface by non-electrical forces called "specific forces", of which Van der Waals forces are the best representation. Let us set:

σ_i: density of counter-charges absorbed to the surface by Van der Waals forces $(C.m^{-2})$

The ensemble thus created is electrically neutral.

$$\sigma_o + \sigma_i + \sigma_d = 0$$

Note, however, that purely Coulombian bond forces would be identical for all ions of the same valence. Such bonds are said to be "indifferent", in opposition to "specific" bonds.

It was the physicist Stern-Hamburg [HAM 24] who introduced "specific" ions adsorbed to the colloidal surface by Van der Waals forces so that, referring to Figure 3 of Stern's publication, the charges and potentials are distributed as shown in our Figure 5.1.

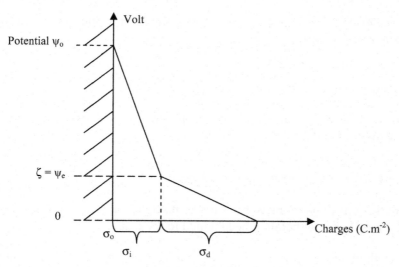

Figure 5.1. *Stern–Hamburg double layer*

Thus, surface ions are:

– non-solvated ions forming an integral part of the surface (layer σ_0);

– solvated ions in contact with the former ions, but not in direct contact with the surface (layer σ_i).

The surface separating σ_i and σ_d is the external Helmholtz surface and the surface separating σ_0 and σ_i is the internal Helmholtz surface.

The potential ψ_e is that of the external Helmholtz surface. It is generally accepted that this potential is the zeta (ζ) potential, which is the potential of the slippage surface of the particle in relation to the liquid under the influence of an electrical field.

If we consider the diffuse layer to be a capacitor with the thickness k, charge σ_d and potential ζ, we can write:

$$\sigma_d = \varepsilon_r \varepsilon_o \zeta \kappa$$

κ: inverse of the Debye length (see section 3.3.2) (m^{-1})

ζ: ζ potential (V)

Hidalgo-Alvarez *et al.* [ALV 85] found the same value by measuring the ζ potential by two different methods.

– sedimentation potential;

– flow potential through a porous substance.

We shall now examine the Stern–Hamburg layer and Gouy–Chapman's diffuse layer in greater detail.

5.1.3. *The Stern–Hamburg rigid layer*

The ions or polymers adsorbed to the particles may be present naturally, or they may have deliberately been added to the dispersion.

The thickness of the adsorbed layer is of the order of the molecular diameter, and it:

– is no greater than 1 nm for ions;

– remains less than 10 nm for polymers, because the free loops and tails whose length is greater than this are not included in the adsorbed layer.

These dimensions are to be compared to the diameter of the particles present in the sols. That diameter is between 10 nm and 1 μm.

The layer of adsorbed species is known as the Stern layer. We can model it as two superficial layers of electrical charges of uniform density distributed across two concentric spheres:

– the inner layer whose density is σ_o, distributed over a sphere whose radius a is that of the particle. That layer corresponds to the natural charge of the particle;

– the outer layer, whose density is σ_a, corresponds to the species adsorbed and distributed over a sphere whose radius is $a + \delta$, where δ is the thickness of the Stern layer.

We shall use the following electrostatic results (see [DUR 53]).

When an electrical charge is distributed over a hollow sphere with a uniform density:

– the field inside the sphere is zero, and the potential is constant;

– the field outside and the potential are the same as if the whole of the charge were concentrated at the center of the sphere.

The charge contained in the sphere delimiting the Stern layer, of thickness δ, is, when viewed from the outside:

$$q_s = q_o + q_a = 4\pi (a + \delta)^2 (\sigma_o + \sigma_a)$$

The Stern layer is surrounded by another layer, called the diffuse layer, where the ions are endowed with the agitation specific to liquids. When the particle moves, it only partially entrains the diffuse layer. The set formed by the particle and the ions or molecules moving along with it is located within a sphere known as the slippage sphere (SS), and it is convenient, and does not conflict with real-world data, to assimilate the SS to the set (particle + Stern layer). The radius of the SS is then equal to $(a + \delta)$.

5.1.4. Diffuse layer: potential profile

Hereafter, we shall assume that the ions obey a Maxwell distribution, where the energy of each ion is $ez_i \varphi$.

We know that the potential φ decreases regularly from the value ζ when r varies from r_I to infinity. The potential φ then tends toward zero. Poisson's differential equation is written as follows, for spherical symmetry.

$$\Delta \varphi = \frac{d^2 \varphi}{dr^2} + \frac{2}{r} \frac{d\varphi}{dr} = \frac{-\rho}{\varepsilon \varepsilon_o}$$

where:

$$\rho = kF\Sigma c_i z_i \exp\left(-\frac{Fz_i\varphi}{RT}\right)$$

r_i: radius of the ion i in view of the adsorbed species

ρ: electrical charge per volume $(C.m^{-3})$

ε_0: permittivity of a vacuum $((36\pi.10^9)^{-1}(C.m^{-1}.V^{-1}))$

ε: relative permittivity of the medium: dimensionless

φ: potential: V

c_i: concentration of the ion i $(kion.m^{-3})$

z_i: algebraic valence of the ion i

R: ideal gas constant $(8314J.K^{-1}.kmol^{-1})$

T: absolute temperature (K)

kF: kFaraday: $96.487 \times 10^6 C.kion^{-1}$

For a given value $r_{i+1/2}$, Poisson's equation takes the form:

$$\varphi'' + a_{i+1/2}\varphi' = y_{i+1/2} \qquad\qquad [5.1]$$

Indeed, we integrate Poisson's equation by a method known as "prediction correction", based on the following values:

$$\varphi_0 = \zeta \quad \text{and} \quad \left(\frac{d\varphi}{dr}\right)_0 = -2\kappa_0\zeta$$

Let us set:

$$\kappa = \left[\frac{e^2}{\varepsilon\varepsilon_0 k_B T}\sum_i N_A c_i z_i^2\right]^{1/2}$$

e: charge on the electron: $1.60219 \times 10^{-19} C$

N_A: Avogadro's number: 6.02217×10^{26} molecules.$kmol^{-1}$

K_B: Boltzmann's constant: 1.38048×10^{-23} $J.K^{-1}$

We vary the value of r by equal increments Δr.

$$r_j = j \Delta r \text{ and } r_{j+1/2} = \left(j + \frac{1}{2} \right) \Delta r$$

In Poisson's equation, written in the form [5.1], we set:

$$a_{j-1/2} = \frac{2}{r_{j+1/2}} \text{ and } y_{j+1/2} = \frac{-F}{\varepsilon \varepsilon_o} \sum_i c_i z_i \exp \left(-\frac{kFz_i \varphi_{j+1/2}}{RT} \right)$$

Poisson's equation can then be integrated between r_i and r_{i+1}.

$$\varphi = \frac{y_{j+1/2}}{a_{j+1/2}} r + B_{j+1/2} e^{-ra_{j+1/2}} + a_{j+1/2}$$

For $r = r_j$, we must have:

$$\varphi = \varphi_i \text{ and } \frac{d\varphi}{dr} = \left(\frac{d\varphi}{dr} \right)_i = \varphi_i'$$

Hence:

$$B_{i+1/2} = \left(\varphi_i' \frac{\varphi_{i+1/2}}{a_{i+1/2}} \right) e^{r_i a_{i+1/2}}$$

$$A_{i+1/2} = \varphi_i - \frac{y_{i+1/2}}{a_{i+1/2}} r_i - B_{i+1/2} e^{-r_i a_{i+1/2}}$$

We do not know the value $\gamma_{i+1/2}$ involved in $c_{i+1/2}$. Thus, we need to calculate by successive iterations, taking the following as the initial value:

$$\varphi_{i+1/2} = \varphi_i$$

Then, we obtain:

$$\varphi_{i+1} = \frac{c_{i+1/2}}{a_{i+1/2}} r_{i+1} + B_{i+1/2} \, \exp\left(-r_{i+1} a_{i+1/2}\right) + A_{i+1/2}$$

and we write:

$$\varphi_{i+1/2} = \frac{1}{2}\left(\varphi_i + \varphi_{i+1}\right)$$

Two or three iterations will suffice.

The total electrical charge contained between the sphere of radius r_I and the sphere of radius r_i is, by integration (trapeze method):

$$q_i = \varepsilon\varepsilon_o \int_{r_I}^{r_i} y 4\pi r^2 dr = \sum_{k=1}^{i} \varepsilon\varepsilon_o y_{k+1/2} 4\pi r_{k+1/2}^2 \Delta r$$

Now suppose that, for a value $r_n = n\Delta r$, the value φ_n is such that:

$$\frac{kF\varphi_n}{RT} < 0.1$$

This means that:

$$\varphi_n < \frac{8\,314 \times 293 \times 0.1}{96.487.10^6} = 0.0025 V$$

In this case, it is possible to write:

$$e^{-x} \# 1 - x$$

Let us set:

$$k_B = \frac{R}{N_A} \quad \text{and} \quad e = \frac{kF}{N_A}$$

Hence:

$$\kappa^2 = \frac{kF^2}{\varepsilon\varepsilon_o RT} \sum_i c_i z_i^2$$

When we bring the ionic strength into play:

$$I = \frac{1}{2}\sum_i c_i z_i^2$$

and, in view of the fact that (electrical neutrality):

$$\sum_i c_i z_i = 0$$

Poisson's equation becomes:

$$\varphi'' + \frac{2}{r}\varphi' = \kappa^2 \varphi$$

We set:

$$\varphi = \frac{u}{r}$$

We obtain:

$$u'' - \kappa^2 u = 0$$

The solution to this equation must tend toward zero as r increases indefinitely. This solution is of the form:

$$\varphi = \frac{A_e^{-\kappa r}}{r}$$

For $r = r_n$, we must have:

$$\varphi_n = \frac{Ae^{-\kappa r_n}}{r_n} \quad \text{Hence} \quad A = r_n \varphi_n e^{\kappa r_n}$$

Thus we have the "tail" of the potential:

$$\varphi = \frac{\varphi_n r_n}{r} \exp\left[-\kappa(r - r_n)\right]$$

We can verify that there is equality between φ'_n stemming from the above numerical calculation and the value of φ' obtained by differentiation of φ.

$$\varphi'_n = \varphi_n \left(\kappa + \frac{1}{r_n}\right)$$

Indeed, on both sides of r_n, we are dealing with the same differential equation.

The electrical charge present between infinity and the sphere of radius r_n is:

$$q_{n\infty} = 4\pi \int_{r_n}^{\infty} \rho r^2 dr$$

The charge density in this domain is as follows, if we refer to Poisson's equation:

$$\rho = \varepsilon\varepsilon_o \kappa^2 \varphi = \varepsilon\varepsilon_o \kappa^2 \varphi_n \frac{rr_n}{r} \exp{-\kappa(r - r_n)}$$

Thus:

$$q_{n\infty} = \varepsilon\varepsilon_o \kappa^2 \varphi_n r_n \int_{r_i}^{\infty} r e^{-\kappa(r - r_n)} dr = \varepsilon\varepsilon_o \kappa^2 \varphi_n r_n I$$

According to Spiegel [SPI 74]:

$$I = \left[\frac{e^{-\kappa(r - r_n)}}{-\kappa}\left(r + \frac{1}{\kappa}\right)\right]_{r_n}^{\infty} = -\frac{1}{\kappa}\left(r_n + \frac{1}{\kappa}\right)$$

Therefore:

$$q_{n\infty} = \varepsilon \varepsilon_o \varphi_n r_n \left(r_n \kappa + 1 \right)$$

We must verify that:

$$q_T = q_s + q_n + q_{n\infty} = 0$$

If this is not the case, we must correct the value of $(d\varphi/dr)_o = \varphi'_o$, for example as follows (linear interpolation or extrapolation):

$$\varphi'^{(3)} = \frac{q_T^{(2)} \varphi'^{(1)} - q_T^{(1)} \varphi'^{(2)}}{q_T^{(2)} - q_T^{(1)}}$$

NOTE.–

It would be useful to replace the above numerical integration with an empirical expression giving φ as a function of r.

NOTE.–

In dispersions of clay in water, the particles are in the form of platelets. Cabane and Hénon [CAB 03] give the integration of Poisson's equation in the case of a flat surface of the particles.

5.1.5. Electrophoresis of colloids

Electrophoresis is the movement, within a liquid, of electrically charged particles under the influence of an electrical field.

Smoluchowski [SMO 03] and Helmholtz [HEL 79] demonstrated that the velocity of a particle is:

$$v_p = \frac{\varepsilon \varepsilon_o \zeta E}{\mu} \qquad \text{(S.I.)}$$

ζ: zeta potential of the double layer surrounding the particle (V)

E: electrical field $(V.m^{-1})$

η: viscosity (Pa.s)

It is understood that, for colloids, the product κa is much higher than 1.

a: radius of the colloid (m)

κ: inverse of the Debye length – i.e. of the thickness of the diffuse layer surrounding the colloid (m^{-1})

$$\kappa^2 = \frac{e^2}{\varepsilon \varepsilon_o kT} \sum_i c_i z_i^2 \qquad \text{(S.I.)}$$

ε_0: absolute permittivity of a vacuum: $8.85434 \times 10^{-12} C.m^{-1}V^{-1}$

k: Boltzmann's constant: $1.38048 \times 10^{-23} J.K^{-1}$

T: absolute temperature (K)

c_i: concentration of the ion i $(ion\text{-}kion.m^{-3})$

z_i: valence of the ion i

ε: relative permittivity of the liquid: dimensionless.

A particle is characterized by its mobility; if v_p is its velocity:

$$u_p = \frac{v_p}{E} = \frac{\varepsilon \varepsilon_o \zeta}{\mu}$$

u_p: mobility of the particle $(m^2.s^{-1}.V^{-1})$

O'Brien and Hunter [OBR 81] put forward an expression of the mobility which is more accurate than Smoluchowski and Helmholtz's, and yet is relatively simple.

5.1.6. *Osmotic pressure of dispersions of colloids*

Readers can refer to the work of Bowen and William [BOW 96].

5.2. Adsorption in the double layer

5.2.1. *Definitions*

1) Adsorption in the rigid Stern layer:

Colloids are considered stable if all the particles carry a charge of the same sign at their surface. Thus, a sol of gold is negatively charged. On the other hand, in the presence of Al^{+++} (20×10^{-6}eq/L), flocculation occurs in 4 hours, because the aluminum ions fix to a portion of the colloids in the rigid layer.

Similarly, at low pH, the H^+ ion affixes to carbon black, whilst at a high pH, it is the OH^- ion which affixes. Between these extremes, there is a value known as the isoelectric point where the charge of the colloids is zero and where the sol is unstable and flocculates very quickly.

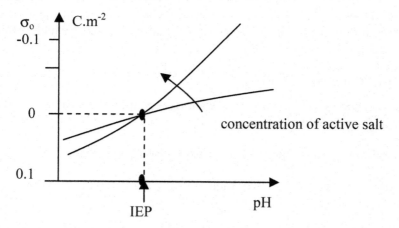

Figure 5.2. *Profile of the variations in surface charge as a function of the pH*

Whatever the concentration of the charge-modifying agent, the isoelectric point (IEP) is common to all the curves.

2) Point of zero charge (PZC):

Metal oxides and -hydroxides can adsorb H^+ and OH^- ions. The surface charge depends on the concentrations of these two types of ions at the surface and, consequently, on the pH of the solution.

The PZC corresponds to a total charge of zero, meaning that there are equal surface concentrations of H^+ and OH^- ions, leaving a neutral balance. These two types of ions are known as the "determinant ions" (of the surface charge).

The "determinant ions" are:

– Ag^+ and I^- for solutions of silver iodide;

– H^+ and OH^- for oxides and hydroxides.

Thus, oxides and hydroxides are charged positively at a low pH and negatively at a high pH.

3) Specifically adsorbed ions:

In addition to the "determinant ions", we may have anions or cations that are also adsorbed depending on the specific characteristics of the crystalline system/solution. These are the "specifically adsorbed ions" which influence σ_d, ζ, and also impact on the surface charge.

4) Isoelectric point (IEP):

The isoelectric point is reached when $\sigma_d = 0$ and $\zeta = 0$. An electrical field has no influence on the particles.

To reach the IEP or the PZC, we first adjust the pH. Hence:

if	$\sigma_o = 0$	then	pH = PZC
if	$\sigma_d = 0$	then	pH = IEP

In the absence of specifically adsorbed ions:

pH = PZC = IEP

In the presence of a specific adsorption:

PZC ≠ IEP

5.2.2. *Variations of the IEP and PZC with adsorption*

Note that the charges in the fixed layer (Stern layer) and the diffuse layer vary in opposite directions.

The pH of the diffuse layer varies in line with that of the solution.

To clarify our understanding, let us look at the result of the specific adsorption of an anion to a surface, such that, initially, IEP = PZC:

To find nullity of the surface charge (that is to say, the PZC), we must have adsorption of H^+ in the Stern fixed layer. In this case, there will be a decrease in the amount of H^+ in the diffuse layer and an increase in the pH values – particularly that of the solution. Hence, the PZC increases with the adsorption of anions.

Adsorption	PZC	IEP
Anion	↗	↘
Cation	↘	↗

Table 5.1.

To find the point of nullity of σ_i and ζ, we shall see that we need to decrease the pH, which enables us to reach the IEP. Indeed, the H^+ ions are attracted by the surface and are therefore less numerous in the diffuse layer. To supply the diffuse layer with H^+ ions, the pH will therefore decrease and we then find the IEP.

5.2.3. *Recap on the influence of pH*

There are practically no charges on molecular crystals (graphite, diamond, orthorhombic sulfur). There are, however, charges carried by ionic crystals.

Surface charges essentially originate in one of two ways:

– either they are created by an electrical imbalance in the crystalline lattice;

– or they result from the interaction between the liquid phase and the surface of the particles. This is specific adsorption.

1) There may be isomorphic substitution in the crystalline lattice. Thus, in clays (which are hydrated aluminosilicates), the Si^{4+} ion may be partly replaced by Al^{3+} and, if the substitution takes place in the vicinity of the surface, we see the appearance of a negative surface charge. Such is the case with kaolin. Zeoliths, which are used as cation exchangers, are another example: the cations are especially mobile in a negatively charged lattice.

2) On contact with water, depending on their pH, certain surface ions react either with H^+ or with OH^-. The reaction product may remain on the surface or pass into solution, depending on its solubility. Thus, at a low pH, the H^+ ion will react with the S^{2-} ion from a sulfide:

$$S^{2-} + H^+ \rightleftarrows HS^- \text{ soluble}$$

The metal cation from the sulfide will remain on the surface, giving it a positive charge.

The same mechanism applies for salts. For example, for a carbonate at low pH:

$$CO_3^{-2} + H^+ \rightleftarrows HCO_3^- \text{ soluble}$$

On contact with water, oxides are hydrated superficially. Thus:

$$O^{2-} + H_2O \rightleftarrows 2OH^-$$

This last reaction does not alter the electrical charge on the surface, but the surface of the oxide is covered with hydroxide $M(OH)_m$. Depending on the pH, the H^+ or OH^- ions will act on the hydroxide. For example, with a trivalent metal:

$$M(H_2O)_n(OH)^{2+} + H^+ \rightleftarrows M(H_2O)_{n+1}^{3+} \qquad \text{(low pH)}$$

or else:

$$M(H_2O)_n(OH)^{2+} + OH^- \rightleftarrows M(H_2O)_n(OH)_2^+ \qquad \text{(high pH)}$$

These reactions also take place in solution.

Particles of organic matter contain amino acids, peptides, proteins, etc., meaning that at their surface, we find amine and acid functional groups. In normal conditions, the acid functional group can relinquish a proton more readily that the amine group can accept it, and the surface is generally negatively charged. Table 5.2 lists the IEP of certain oxides.

Oxide	SiO_2	TiO_2	Fe_2O_3	Al_2O_3	MgO
IEP	2	6	8	9	12

Table 5.2. *Isoelectric points*

NOTE.–

Usually, the following ions are present in the water used for industry:

– Na^+ (for softened water);

– Ca^{++} (for hard water);

– Mg^{++} (generally in a much smaller proportion than that of Ca^{++});

– PO_4^{---} (in treated water for boilers);

– SiO_3^{--} (water in contact with clay-rich terrain);

– CO_3^{--} and HCO_3^- (presence of dissolved carbon dioxide),

– OH^- (presence of a basic pH).

The analytical methods often consist of neutralizing the solution using caustic soda or dilute sulfuric acid in the presence of a colored indicator. We shall not go into detail about these methods.

Thus, before turning our attentions to complicated treatments, we must check whether or not a moderate variation in the pH is sufficient to destabilize the dispersion.

5.2.4. *Specific adsorption of ions at the surface of the particles*

The liquid phase may contain certain ions which, depending on their nature, may be adsorbed and held in place by various types of bonds. We can cite:

– pure or partial ionic bonds;

– hydrogen bonds;

– Van der Waals bonds;

– bonds formed during a genuine chemical reaction, whose products may partially or completely pass into solution.

The last 3 bonds correspond to specific adsorption.

If the bond is purely ionic, we see the exact neutralization of a natural charge by the adsorbed charge, but this is an exceptional case. In general, the bond established is independent of the charges carried by the atoms of the ion, which are not directly affected by the adsorption. These charges remain free and, across the whole of a particle, are added to the natural charge. The overall charge of the particle after adsorption is:

$$q_s = q_o + q_a$$

q_o: natural charge

q_a: total charge of adsorbed ions

We have just seen that there is no automatic link between q_o and q_a. By appropriately choosing the nature and concentration of the adsorbed ions, it is possible to render q_s equal to zero. It is sometimes said that we have reached the point of charge reversal (PCR). In practical terms, to destabilize a sol, it is not always necessary to cancel out q_s; it may be enough that the absolute value of that charge is sufficiently low. We shall see that in terms of the study of the frequency of clumping of particles.

Let us examine a few examples of adsorption.

Monovalent ions such as Na^+ and K^+ establish a simple ionic bond with the negative charge with which they are faced.

Organic particles of natural water contain such adsorbed ions. Similarly, kaolinite, which is also negative, is surrounded by the same ions, in addition to which the Ca^{2+} and Mg^{2+} ions are present. This type of adsorption is insufficient, by far, to neutralize the charge of the particles.

Bivalent ions are, generally, more actively adsorbed than monovalent ions. Thus, Ca^{2+} affixes to the surface of particles of MnO_2. The Ba^{2+} ion is adsorbed to Fe_2O_3, and can lead to a charge reversal.

Trivalent ions (or even quadrivalent ones) and of low diameter are strongly solvated by water (e.g. Al^{3+} and Th^{4+}). When they are adsorbed, the water molecules prevent too close an approach to the surface, so that the charge involved in the bond is much less than the total charge of the ion. In this case, there may be a charge reversal.

For certain conditions of pH, trivalent ions such as Fe^{3+} and Al^{3+} give rise to the formation of complexes of variable charge depending on the complex formed, which are adsorbed by a hydrogen bond to the oxygen in an acidic functional group.

Polyphosphate ions contribute negative charges to the positive particles of Al_2O_3, and certain quaternary ammonium ions contribute positive charges to negative organic particles.

5.2.5. Adsorption of polymers: flocculation

The chain of a molecule of polymer is a succession of patterns which may carry positive or negative electrical charges, or indeed carry no charges whatsoever. We distinguish between:

– anionic polymers, charged negatively;

– cationic polymers, charged positively;

– non-ionic polymers, which have no charge.

The first two categories are known as polyelectrolytes. This term does not apply to the third.

After adsorption of a polymer, clumping of the particles may occur:

– either by electrostatic attraction (zone mechanism);

– or by bridging.

For bridging between two particles to be facilitated, the ideal would be that the absorbed polymer resemble the spines of a sea urchin. During bridging, the spines of two nearby urchins interpenetrate with one another, and the end of the spines of one becomes affixed (adsorbed) to the surface of the other. In order for the comparison to be more realistic, we must accept that those spines have a certain degree of flexibility, rather than simply being rigid.

In order to resemble the spines of an urchin, the polymer molecules need to be as rectilinear as possible in solution before adsorption. This will be the case if the liquid phase possesses the following properties:

– the solvent has enough of an affinity for the polymer (otherwise the molecule rolls up into a ball);

– the pH of the solution imbues the successive patterns of the chain with electrical charges, which repel one another;

– the ionic strength of the solution is moderate. Indeed, if it is too high, the counter-ions in the solution partially neutralize and "screen" the charges in the chain.

The above criteria must be slightly adjusted, because the polymer molecule must not be rigid, but retain a certain flexibility. Indeed, adsorption could not only happen at one end of the chain, because the bond thus established would be too weak. In reality, adsorption affects patterns spaced along the length of the chain and, as the molecule is flexible, there are free loops between the fixed patterns, and also a tail at each end of the molecule. It is the tails and loops which act as the urchin spines.

Even though, on each individual pattern adsorbed, the bond is of mediocre solidity, the multiplicity of those bonds means that the adsorption of the whole molecule is solid and irreversible. Of course, the affinity of each pattern for the surface of the particle must be greater than its affinity for the solvent.

In practice, the fixed mass is, at most, $1-2$ mg.m^{-2} for a monomolecular layer. The patterns adsorbed represent only $0.3-0.5$ mg.m^{-2}, and the rest remains in loops and tails.

Adsorption is a slow phenomenon, and time is an important factor. If we allow the adsorption of a molecule of polymer to continue indefinitely, without the intervention of other phenomena, the molecule will eventually flatten itself totally against the solid surface, and the spines of the urchin will have gradually disappeared after progressively becoming shorter and shorter.

One might be forgiven for thinking that if we use a solution that is sufficiently concentrated in terms of polymer, the adsorbed molecules will be numerous enough to bring about competition for adsorption sites. Indeed, the loops and tails would remain, but there would no longer be sites available on the urchin for the adsorption of the ends of the spines of a neighboring urchin. Thus, we would eventually see the formation of a protective layer of polymer over the particle.

However, generally we use polymers whose electrical charge is *opposite* to that of the particle. As we saw in the case of ions, adsorption may then lead to a localized charge reversal at the site of the molecule of polymer. In addition, if the concentration of polymer is insufficient to cover the whole of the surface of the particles even after a very long time, there will be zones where the sign of the charge is different and, if the particles coming together are correctly orientated, we will see an electrostatic attraction between zones whose charge is of opposite sign.

However, this mechanism is not common, because we need to account for the destabilizations caused by non-ionic polymers (and also by numerous polyelectrolytes). This can only be done through the bridging model.

As time progresses, there is competition on the surface of the particles between:

– homoadsorption, which is the fixation of additional patterns belonging to the molecules already adsorbed;

– heteroadsorption, which is the fixation of patterns belonging to molecules adsorbed on another particle. This is, strictly speaking, the mechanism of bridging.

Whilst homoadsorption inevitably occurs very soon after the addition of polymer to the liquid phase, we must, without delay, implement the appropriate mechanical means to encourage heteroadsorption.

From the above, it results that, if we gradually increase the polymer content of the dispersion, destabilization eventually occurs, but then if the content continues to rise, the dispersion becomes stable once again, because:

– either the surface of the particles is completely covered with zones having undergone charge reversal;

– or else there are no more sites available for heteroadsorption.

Besides the time and the concentration, the molecular mass of the polymer used has a significant influence on its properties. Thus, polyethylene oxide is a stabilizing agent if its molecular mass is moderate and has a destabilizing influence if its molecular mass increases. Indeed, it is a non-ionic adjuvant which is certain to act by bridging, and this mechanism is all the more efficient when the urchin spines are long.

Note, in closing, that a polyacrylamide with molecular mass equal to 10^6 has a chain length equal to 0.1 μm once deployed. This value of molecular mass is the minimal threshold beyond which bridging is possible.

5.3. Interaction between two particles

5.3.1. Hamaker's law [HAM 37]

The force of *attraction* between two particles (supposed to be spherical and identical) is:

$$F = -\frac{A}{24x^2 d_p} \quad \text{where} \quad x = \frac{e}{d_p} \ll 1$$

d_p: diameter of the particles (m)

e: gap between the surfaces of the two particles (m)

Very generally, e is greater than or equal to 50 nm. The constant A, according to Hamaker, is such that:

$$10^{-20} J < A < 10^{-19} J$$

Note that Hamaker's law is merely a practical application of the behavior of the Van der Waals forces.

Values of the Hamaker constants can be found in Lyklema [LYK 68] and in Visser [VIS 72]. In the absence of data, Hamaker proposes an average value:

$$A = 0.7 \times 10^{-19} J$$

The potential bond energy is:

$$E = \frac{A}{24x} \qquad (J)$$

EXAMPLE.–

$e = 50$ nm $\qquad A = 0.7 \times 10^{-19} \qquad d_p = 20$ μm

$$F = -\frac{0.7.10^{-19}}{24 \times 20.10^{-6}} \left(\frac{20.10^{-6}}{50.10^{-9}} \right)^2 = 2.33.10^{-11} N$$

$$E = -\frac{0.7.10^{-19}}{24} \times \left(\frac{20.10^{-6}}{50.10^{-9}} \right)^2 = 1.17.10^{-18} J$$

In order for Hamaker's law to be meaningful, the following conditions must be met:

– the medium is calm – i.e. there is no agitation;

– the particles are small ($d_p < 50$μm);

– the surface charge of the particles is zero;

– the liquid has a low viscosity, otherwise the particles will have difficulty in coming together.

In the case of a crystallization operation, it is good for the oversaturation to be high ($\sigma > 0.2$, see section 2.1.1 in [DUR 16d] for the definition of σ).

5.3.2. *Attraction and repulsion between two particles supposed to be spherical and identical*

According to Derjaguin and Landau [DER 41], and to Vervey and Overberk [VER 48], the energy of interaction E between two identical spherical particles is the sum of a negative term of attraction [HAM 37] and a repulsion term:

$$E = -\frac{Aa}{12\,h} + 4\pi\varepsilon_2\varepsilon_o \frac{a^2}{h+2a}\,\varphi_o^2 \exp(-\kappa h)$$

φ_o: surface potential (Volt)

A: Hamaker constant (J)

ε_o: permittivity of a vacuum (C.V^{-1}.m^{-1})

ε: relative permittivity of the liquid

a: radius of the spheres (m)

h: distance between the surfaces of the two spheres (m)

k: Boltzmann's constant (1.38048×10^{-23}J.K^{-1})

T: absolute temperature (K)

$1/\kappa$: Debye length (m)

As we saw in section 3.3.2 in [DUR 16a], $\kappa^2 = \dfrac{e^2}{\varepsilon_2\varepsilon_o kT}\displaystyle\sum_i c_i z_i^2$

e: electronic charge (1.60206×10^{-19}C)

c_i: ion concentration (kion.m^{-3})

z_i: valence of the ion i

From the point of view of the units, note that the equation of the plate capacitor equivalent to the diffuse layer is:

$$\sigma_d = -(\sigma_o + \sigma_i) = \varepsilon_r\varepsilon_o\kappa\varphi_o \qquad \text{(S.I.)}$$

The plateau AB corresponds to the fact that the two particles can come no closer together than AB. The plateau AB is a zone of stability, and this is due to the impediment caused by the molecules of solvent.

At C, there is an energy barrier which must be crossed in order for two separate particles to be able to come together.

At D, there is a potential minimum, which corresponds to a "secondary" coagulation, but the corresponding clumps are not very strong, and are easily dispersed.

On the other hand, the clumps corresponding to the plateau AB – i.e. to the primary minimum – are irreversible.

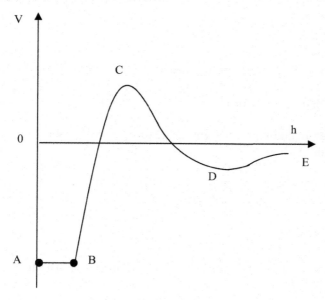

Figure 5.3. *Energy of interaction between two particles*

NOTE.–

We know that the ionic strength is defined by:

$$I = \frac{1}{2}\sum_i c_i z_i^2$$

When we increase the ionic strength by the use of trivalent ions in the solution such as Fe^{+++} or Al^{+++}, we greatly decrease the Debye length $1/\kappa$. However, that length is exactly the thickness of the diffusion layer – that is, the particles can approach one another and not repel each other until the point where the attraction term predominates, and we then see the coagulation of the colloidal dispersion.

5.3.3. Coagulation of sols according to Ostwald [OST 35, OST 39]

Ostwald studied dispersions of colloids (which Jean Perrin called "sols"). Certain colloids are positively charged, and others negatively:

– negative charge: As_2, S_z, Sb_2S_3, MnO_2, Congo Red, Gold (metal), Silver (metal), Platinum (metal), sulfur;

– positive charge: Fe_2O_3, Al_2O_3.

Coagulation occurs when, to such a sol, we add a neutral electrolyte in salt form. The active principle of that electrolyte is the ion whose charge is opposite to that of the colloid.

Coagulation takes place after a latency period, which may be between 5 seconds and 24 hours.

Ostwald evaluates the activity coefficient of the active ion using a simplified formula (but an accurate one, when the concentration is low).

$$\log_{10} \gamma_i = -0.5\, z_i^2 \sqrt{I / n_i}$$

I is the ionic strength for the solution of a single electrolyte.

$$I = \frac{1}{2}\left(c_+ z_+^2 + c_- z_-^2\right)$$

c: concentration ($kmol.m^{-3}$)

n_i: number of ions i in the molecule

Coagulation happens when the activity coefficient of the active ion reaches a precise value, depending on the colloid. The values are to be found in Ostwald [OST 35].

Armed with these results, we are able to switch directly from one electrolyte to another without the need for any experimentation.

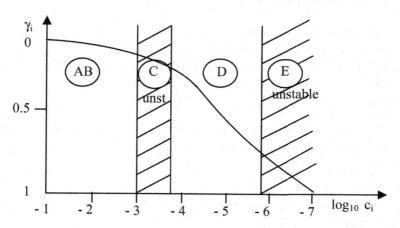

Figure 5.4. *Coagulation zones (from [OST 39])*

NOTE.–

In his work, Ostwald [OST 39] confirmed what has been said regarding "primary coagulation" and "secondary coagulation", as shown by his Figure 11, from which our Figure 5.4 is taken.

Supposing that the colloid is initially negative (zone AB), if we increase the concentration of the active (positive) ion, we see (using the notations in Figure 5.4):

– a zone of instability of coagulation (zone C);

– a zone of stability (zone D). This stability of the coagulate is said to be secondary;

– a new zone of instability (zone E).

– zone E corresponds to a slight contraction of the double layer, and corresponds to what we call secondary coagulation. The zone AB corresponds to primary coagulation.

5.4. Practice of destabilization

5.4.1. *Destabilization methods*

The principle of such methods is to encourage the coming together and clumping of the particles, because we then obtain clumps which are much larger in size, which can easily be separated from the liquid phase by decantation or filtration. To do so, as we have seen, we can employ coagulation or flocculation.

The sediment or filtration cake obtained by coagulation is called a coagulate. The same solid phase obtained by flocculation is called a flocculate.

5.4.2. *Gibbs energy of interaction of two particles; critical temperature*

The sol will be less stable when the particles can more easily approach one another. The Gibbs energy of interaction of a couple of identical particles is:

$$G_T = G_A + G_R + G_S$$

We know that a spontaneous phenomenon corresponds to a decrease in Gibbs energy. This is what must happen as the particles approach one another. In reality, we shall see that G_T corresponds to an energy barrier which must be crossed, which we shall take into account when studying the kinetics of particle clumping.

G_A expresses the mutual attraction, and stems from the Van der Waals forces (see section 5.3.1).

G_R represents the electrostatic repulsion and, by simplifying, we can write:

$$G_R = \frac{Cq_s^2}{h + 2(a + \delta)} \exp(-\kappa h)$$

q_s: charge of a slip sphere SS (Coulomb)

C: constant which increases with the particle size

Readers can also refer to section 5.3.2.

It will be easier for the particles to come together if G_R is small, which corresponds to a high value of κ, so to thin diffuse layers. We saw earlier that it is possible to obtain this result by increasing the ionic strength of the solution (contraction of the diffuse layer).

In addition, we can immediately see from the expression of G_R that this repulsion term shrinks rapidly if the charge q_s of the SS tends toward zero. We obtain this result by adsorption of ions with the opposite charge of the natural charge of the particles.

G_S is a steric term which comes into play when a polymer is adsorbed to the particles. The particles are encouraged to come together if it corresponds to:

$$\Delta G_T < 0$$

However, we can write:

$$\Delta G_T = \Delta H - T\Delta S$$

Statistical thermodynamics shows that the compression of a polymer chain corresponds to a decrease in entropy. Thus, if two layers of polymers coming into contact with one another compress each other elastically and athermically, we have:

$$\Delta G_T = -T\Delta S > 0$$

Rapprochement, then, will be prevented (more specifically, in this case, ΔS corresponds to the variation in entropy of configuration of the polymer chains).

Note that electrostatically charged chains repel one another, but that an increase in the ionic strength of the solution partially "masks" the charges and helps encourage them to come closer together. Thus, the ionic strength decreases the effect of elastic compression.

This thermodynamic formalism can also be used to specify the influence of the solvent's affinity for the polymer. Suppose that the liquid is a good solvent, and in particular, that dissolution results in:

– the release of heat: $\Delta H < 0$

– the solvation of the polymer, i.e. a higher-order degree: $\Delta S < 0$.

If two particles come together, the interstitial solvent is expelled and the result is the opposite of a dissolution. Hence:

$$\Delta H > 0 \quad \text{and} \quad \Delta S > 0$$

Depending on the temperature, we may have:

$$\Delta G_S = \Delta H - T\Delta S \qquad \text{(positive or negative)}$$

The critical temperature, called the θ temperature, is a value above which the presence of the polymer favors destabilization and below which the sol is stable. Conversely, if we are dealing with a poor solvent in the sense described above, destabilization will occur by cooling.

In concrete terms, if the temperature is such that the solvent has affinity for the polymer, the layers will not mix and the rapprochement will be hampered (and if the liquid has low dissolving power, the opposite is true).

5.4.3. *Frequency of clumping and energy barrier*

At a high frequency of clumping, we see the rapid destabilization of the dispersion. On the other hand, if the frequency of clumping is very low, the dispersion is practically stable.

The frequency of clumping between the particles or the clump of types 1 and 2 is expressed by:

$$f = e^{\frac{-E}{k_B T}}.k\, n_1.n_2$$

f: frequency $(m^{-3}.s^{-1})$

E: energy barrier, which we shall assimilate to the Gibbs energy G_T seen in the previous section (J)

k_B: Boltzmann's constant $(J.K^{-1})$

T: absolute temperature (K)

k: constant characteristic of the phenomena causing the collisions $(m^3.s^{-1})$

n_1 and n_2: number of particles or of type 1 and type 2 clumps per m^3 of the dispersion (m^{-3})

The energy E is the barrier that needs to be crossed in order for two slip spheres to come into contact with one another. As we have seen, this energy includes three terms:

– Van der Waals attraction;

– electrostatic repulsion of the two double layers at the moment when the diffuse layers interpenetrate;

– interaction of any adsorbed polymer layers.

The following conclusions can be drawn from the results established above:

1) The dependence of the Van der Waals attraction depends on the nature of the substances present and is unpredictable. That shows that it is not possible to have an action on that term.

2) Experience shows us that the electrostatic repulsion of the two double layers decrease if:

– the thickness of the diffuse layers decreases, and to obtain this result, it is helpful to increase the ionic strength of the solution;

– the total charge q_s of the SS (slip sphere) decreases. It is then useful to adsorb to the particles, species with a charge whose sign is opposite to that of the natural charges, though of course without surpassing the point of charge reversal. In practical terms, we see that the clumping of the particles is no longer impossible when the ζ potential is less than 20 mV, which corresponds to a charge that is proportional to the radius of the SS. More specifically, in water, we obtain:

$$q_s = 1.78 \times 10^{-10} (a + \delta)$$

q_s: Coulomb

$a + \delta$: radius of the SS (m)

For a particle of 0.1 μm, in the knowledge that the charge on the electron is 1.6×10^{-19} C,

$$q_s = \frac{1.78\times10^{-10}\times10^{-7}}{1.6\times10^{-19}} = 110 \text{ electron charges}$$

According to this criterion, a 1 nm molecule should have a charge smaller than that carried by an electron.

3) We have seen that the adsorption of a polymer and its consequences for the energy barrier are complex. We shall simply recap the following conclusions:

– the concentration of the polymer must be within a narrow range;

– the polymer's affinity for the solvent must be moderate;

– it is helpful for the electrical charge of the polyelectrolyte to be opposite to the particle's natural charge;

– the molecular mass of the polymer must generally be greater than 10^6 kg.kmol^{-1} if it acts by bridging.

5.4.4. Collision mechanisms

In the expression of the clumping frequency f, the coefficient k characterizes the conditions in which the collisions take place. At this point, let us distinguish between two common cases:

– clumping takes place in a vat with gentle stirring:

$k = k_P + k_O$

– clumping takes place in a calm environment in a decanter:

$k = k_P + k_D$

k_P: perikinetic collision constant

k_O: orthokinetic collision constant

k_D: constant of collisions due to differential decantation.

Let us examine these three mechanisms:

1) Perikinetic collisions:

The cause of the collisions, here, is the agitation due to Brownian motion:

$$k_P = \frac{2}{3} \frac{k_B T}{\mu} \frac{(d_1 + d_2)^2}{d_1 d_2}$$

d_1 and d_2: diameters of the clumps or of the SS of types 1 and 2 (m)

μ: viscosity of the liquid phase (Pa.s)

If one of the particles is significantly smaller than the other, i.e. if:

$$\frac{d_2}{d_1} = r \ll 1$$

then:

$$k_P = \frac{2}{3} \frac{k_B T}{\mu} \frac{1}{r}$$

If the two clumps or particles are of equal sizes, then the constant k is independent of their common size:

$$k_P = \frac{8}{3} \frac{k_B T}{\mu}$$

Let us, for instance, consider the situation at 20°C (293K) in water ($\mu = 10^{-3}$ Pa.s) and, knowing that Boltzmann's constant is equal to 1.38×10^{-23} J.K^{-1}, we then obtain:

$$k_P = 1.08 \times 10^{-17} m^3.s^{-1}$$

2) Orthokinetic collisions:

If, because of mechanical agitation, the instantaneous velocity of the fluid varies in a direction perpendicular to its normal motion (presence of a velocity gradient equal to G), the resulting shearing motion causes collisions:

$$k_O = \frac{G(d_1 + d_2)^2}{6}$$

We can show, in fluid mechanics, that the power P/V dissipated per unit volume of a fluid in a laminar flow regime is a quadratic function of the velocity gradients multiplied by the viscosity.

It is therefore logical to write:

$$G = \left[\frac{P}{\mu V}\right]^{1/2}$$

During an operation of particle clumping through orthokinetic collisions, the agitation must be very gentle so as not to destroy the clumps formed, or any polymer chains which may be present. The typical value of the ratio P/V is no greater than $1\,W.m^{-3}$, and this figure can be compared to the volumetric power expended for a chemical reaction in a homogeneous medium, which is around $500\,W.m^{-3}$. If, in addition, we accept that the viscosity of the liquid is equal to that of water – i.e. 10^{-3} Pa.s – we obtain:

$$G = \left[\frac{1}{10^{-3}}\right]^{1/2} \#\,30\,s^{-1}$$

Let us also accept that the two particles have a shared size d_p, and we obtain:

$$k_O = 30 \times \frac{2}{3} d_p^2 = 20\ d_p^2$$

3) Collisions due to differential decantation:

If two particles of different sizes decant in a liquid, the larger will move faster than the smaller, and if the larger one starts at a higher level near to the smaller, the twain shall meet. In this case:

$$k_D = \frac{\pi \Delta \rho G}{72*} (d_1 + d_2)^3 (d_1 - d_2)$$ [5.3]

$\Delta \rho$: difference between densities of the solid phase and the liquid (kg.m^{-3})

g: acceleration due to gravity: 9.81 m.s^{-2}

If:

$$\frac{d_2}{d_1} = r \ll 1$$

then:

$$k_D = \frac{\pi \Delta \rho g d_1^4}{72 \mu}$$

If the large particle is actually a clump, it contains interstitial liquid between the particles making it up. Thus, its density is intermediary between that of the solid phase and that of the liquid. The expression of k_D [5.3], therefore, is an upper bound.

To gain maximum benefit from differential decantation, we must be careful to eliminate the large particles before the destabilization operation.

4) In 1917, Smoluchowski proposed a value of the coefficient k, for which we shall now give a simplified justification.

We shall work on the basis of the hypothesis that each clump is the center of a "sphere of influence", whose radius is R. Two clumps are likely to combine if the distance between their centers is 2R.

At a given moment, the concentration of clumps containing i elementary particles is c_i.

Consider a clump that is the center of a sphere of radius 2R. Over the surface of that sphere, the clumps are distributed in accordance with their concentrations c_i.

If the clump flux density moving toward the center clump is ϕ, we have the following (because of Fick's law):

$$\phi_i = -D\frac{\delta c_i}{\delta r}\# + \frac{Dc_i}{R}$$

The flux k of clumps moving toward the center clump is:

$$\pi(2R)^2 \frac{Dc_i}{R} = 4\pi R D c_i$$

If the center clump is of the type (j − i), the meeting of such a clump with a clump i will yield a clump of type j:

$$\frac{\delta c_j}{\delta \tau} = 4\pi R D c_i$$

The concentration of clumps of type j − i is c_{j-i}. The number of collisions giving rise to a type j clump is, per unit time and volume:

$$\frac{dc_j}{d\tau} = 4\pi R D c_i c_{j-i} = k c_i c_{j-i}$$

5.4.5. *Predominant collision mechanism*

In industry, clumping essentially takes place in two ways:

– either in a tank with gentle agitation. In such a tank, even when agitation is only gentle, decantation is negligible. This operation occurs before filtration and, to be separated in this manner, the clumps must be larger than 10 μm;

– or in a gravity decanter. In order for decantation preceded by clumping to be financially viable, the clumps must be larger than 100 μm. In a gravity decanter, obviously, there is no agitation.

Hereinafter, we shall be working with numerical results obtained using the following data:

$$G = 30\,s^{-1} \qquad\qquad \mu = 10^{-3}\,Pa.s \qquad\qquad \Delta\rho = 1000\;kg.m^{-3}$$

$$T = 293\;K \qquad\qquad\qquad\qquad\qquad k_B = 1.38\times10^{-23}\,J.K^{-1}$$

1) Stirred tank:

– Orthokinetic collisions:

d_P	k_O
10 nm	2.10^{-15}
1 μm	2.10^{-11}

– Perikinetic collisions:

For particles of the same size:

$$k_P = 1.08\times10^{-17}\,m^3.s^{-1}$$

We can see that, in a treatment tank, the effect of Brownian motion is predominant for particles or clumps smaller than 1 μm.

2) Gravity decanter:

– Differential decantation:

If: $r \ll 1$ and $d_1 = 100$ μm, then:

$$k_D = 4.3\times10^{-11}\,m^3.s^{-1}$$

– Perikinetic collisions:

R	k_P
1	1.08×10^{-17}

0.01 2.7×10^{-16}

10^{-4} 2.7×10^{-14}

In a gravity decanter, differential decantation is predominant.

5.4.6. *Destabilization by trapping*

Let us recap the expression of the clumping frequency:

$$F = e^{\frac{-E}{k_B T}} k\, n_1 n_2$$

We can see that, roughly, this frequency is proportional to the square of the concentration of solid particles and, if that concentration is low, as is the case in the treatment of most wastewater, destabilization will take a long time. Therefore, we use the trapping method.

The adjuvant added to the dispersion reacts with the liquid phase and forms a precipitate, which decants of its own accord. The particles initially present in the suspension are trapped in the precipitated and are entrained by it. Thus, we are not dealing with clumping in the true sense. However, the purely mechanical disturbances created by the precipitate in the medium increase the frequency of the collisions. This is a secondary effect.

For this, we use iron sulfate or aluminum sulfate, and we set the pH of the liquor high enough for the hydroxide to precipitate.

5.4.7. *Recap of the distinction between coagulation and flocculation*

Leaving aside the destabilizations obtained solely by acting on the pH, we can ultimately state that:

1) With regard to coagulation by acting on the ionic strength, destabilization occurs beyond a certain minimum concentration of adjuvant, and persists if we increase that concentration. The concentration of solid particles has no bearing on the phenomenon. The ions used as adjuvants which act in this way are known as indifferent ions.

2) As regards flocculation, this destabilization occurs only in a relatively narrow range of concentration of adjuvant. The upper limit of concentration is determined by the appearance of phenomena such as charge reversal or a protective polymer layer. The optimal concentration of adjuvant is proportional to the surface of the solid phase present in the dispersion.

5.4.8. Smoluchowski kinetics [SMO 17]

The disappearance of clumps with index j is written as:

$$kc_j \sum_i c_i \quad \text{where} \quad \sum_i c_i = c_1 + c_2 + \ldots\ldots\ldots + c_n$$

The appearance of pairs is:

$$k\left(c_1 c_{i-1} + c_2 c_{i-2} + \ldots\ldots + c_{i-1} c_1\right)$$

Thus, finally:

$$\frac{1}{k}\frac{dc_k}{dt} = \left(c_1 + c_{i-1} + \ldots\ldots\ldots + c_{i-1} c_1\right) - 2c_i \sum c$$

EXAMPLE.–

$$\frac{1}{k}\frac{dc_1}{dt} = -2c_1 \sum_1^{\infty} c_i$$

$$\frac{1}{k}\frac{dc_2}{dt} = c_1^2 - 2c_2 \sum c$$

$$\frac{1}{k}\frac{dc_3}{dt} = 2c_1 c_2 - 2c_3 \sum c$$

$$\frac{1}{k}\frac{dc_i}{dt} = \left(c_1 c_{i-1} + \ldots\ldots\ldots + c_{i-1} c_1\right) - 2c_i \sum c$$

The author gives the solution to this system (without detailing the method).

$$c_1 = \frac{c_o}{\left(1 + kc_o t\right)^2}$$

$$c_2 = c_o \frac{kc_o t}{\left(1 + kc_o t\right)^3}$$

$$c_k = c_o \frac{\left(kc_o t\right)^{i-1}}{\left(1 + kc_o t\right)^{i+1}}$$

Let us set:

$$T = 1/\left(2\pi DR c_o\right)$$

The maximum of c_k is obtained after time:

$$t_{Max} = \left(\frac{i-1}{2}\right)T$$

and that maximum is:

$$c_k = 4c_o \frac{\left(i-1\right)^{k-1}}{\left(i+1\right)^{k+1}}$$

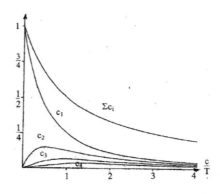

Figure 5.5. *Evaluation of the concentrations of clumps over time*

NOTE.–

The presentation above is simpler than Smoluchowski's, but achieves the same result, where:

$k = 4\pi DR$

This is unimportant, as R is arbitrary, and only the coefficient k needs to be close to reality.

NOTE.–

Using the notion of kinetics of a chemical reaction, Gardner and Theis [GAR 96] claim to express the evolution over time of the complete distribution of clump sizes, but the calculations are much more complicated.

In practice, the coagulation rate cannot be calculated, and is found through experimentation. However, interested readers may refer to the following publications:

– Smoluchowski (1917) [SMO 17]

– Hogg R. *et al.* (1966) [HOG 66]

– Gardner K.H. *et al.* (1996) [GAR 96]

5.5. Destabilizing adjuvants

5.5.1. *Ions as adjuvants*

The most common adjuvant ions are aluminum sulfate, aluminum polychloride, ferric sulfate and calcium in the form of lime.

The form taken by the action of the iron depends on the pH:

– at very low pH, the ion Fe^{3+} acts as an indifferent ion;

– at a low pH, iron acts by adsorption in the form of the complex ion $Fe\,(H_2O)_6^{3+}$;

– at a pH of between 4 and 7, iron precipitates in the form of hydroxide, and acts by trapping.

The action of aluminum is similar to that of iron, but the proportion of positive hexahydrate complex is much greater than in the case of iron, which accounts for the preference of aluminum salts.

The concentration used in wastewater treatment varies from a few grams to several tens of grams per m^3. In the ore industry, lime is used at concentrations of up to 200 g per tonne of solid material.

5.5.2. Polymer adjuvants

Polymers may be natural or synthetic.

1) Natural polymers:

The best-known natural polymers are:

– guar gum, which is derived from the fruit of the carob tree, and is a polysaccharide. It is non-ionic;

– gelatin, which is widely used in the mineral industry. Gelatin is both anionic and cationic, and is adsorbed by a hydrogen bond. It destabilizes more or less any fine dispersion;

– starch, in pre-gelatinized form and therefore water-soluble (corn starch or potato fecula). We can find non-ionic starches, anionic oxidized starches and cationic starches, treated with amines;

– sodium alginate, which is derived from algae, and is anionic. It is very widely used in France, for instance, for drinking water, but has the disadvantage of being precipitated by hard water.

2) Synthetic polymers:

Synthetic polymers tend, above all, to be polyacrylamides. Certain derivatives have patterns that are at once weakly acid and weakly basic (carboxylic acid and tertiary ammonium). Others have strongly acidic or strongly basic functional groups (sulfonic acid and quaternary ammonium).

Such polymers are characterized by:

– the molecular mass;

– the ratio of strength of the acid group to the strength of the basic group;

– the distribution of active functional groups along the chain.

Acrylamide monomer is a toxic product which accumulates in the organism, and it is a barrier to the use of the corresponding polymer in drinking-water treatment.

3) Conclusion:

At equal concentrations, natural polymers are less effective than synthetic ones, although the former are significantly less expensive. Currently, we are seeing a trend toward the substitution of synthetic products for natural products.

The order of magnitude of the concentrations used is $0.1-1 g.m^{-3}$ for water treatment and a few grams per tonne of solid in the ore industry.

Flotation: Froth Columns

6.1. Flotation

6.1.1. *Principle of flotation*

Flotation is a method for separating solid grains of different natures, by catching hydrophobic grains with air bubbles. These loaded bubbles then come together at the top of the device, forming a froth.

Hydrophilic grains do not attach to the bubbles, and remain in suspension in water. This suspension is known as *pulp*. The solid from the froth is called concentrate (after bursting bubbles). In general, the concentrate is the ore we wish to recover.

In order for this separation to be possible, the solid phase must first have been milled to the release size, where each particle is homogeneous and is composed either of pure ore, or of gangue (i.e. waste material). The gangue is generally hydrophilic.

It stems, from the above, that the volume of a flotation cell contains three zones. From bottom to top, we have:

– the zone of air dispersion, in the form of fine bubbles;

– the zone of flotation in the true sense (or capture zone), where the bubbles affix to and capture the solid particles. This zone is naturally the largest in terms of volume and height;

– the zone of froth. The aforesaid froth is evacuated by overflow, assisted by scrapers.

The quantity of ore remaining in the pulp after exiting the flotation cell is called the "lost fraction", and not, the "tails" (as a clumsy person would say).

Trahar and Warren [TRA 76] give a qualitative description of the mechanism of flotation.

6.1.2. *Fixation between bubbles and ore particles*

For the flotation operation to be feasible, it is necessary for permanent contact to be maintained between an air bubble and a particle of ore.

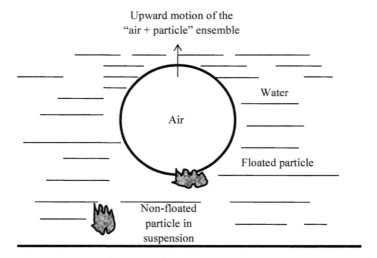

Figure 6.1. *Principle of flotation*

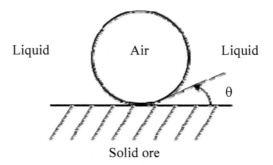

Figure 6.2. *Contact between ore and air bubble*

In the technique of flotation, the contact angle θ is measured in the liquid so that (see Figure 6.2):

– if θ tends toward zero, the contact (and therefore the adherence of the air bubble to the ore) tends to reduce to nothing;

– if θ tends toward π radians, then the ore tends to affix solidly to the bubble.

Rather than directly measuring the contact angle, Wark and Cox [WAR 35] looked for the values of the parameters (composition of the aqueous phase) where an air bubble remains attached to the surface of the ore, and developed a device for this purpose.

They noted, to begin with, that sodium cyanide (which is very widely used in mineralurgy) decreases fixation, and secondly, that fixation depends heavily on the pH. They established what they call contact curves which, for each ore, divide the plane whose abscissa is the pH and ordinate the cyanide concentration into two domains.

If, in addition, we conduct tests with an adjuvant (a collector, a depressor, etc.), we obtain a web of curves, with each curve corresponding to a given concentration of the adjuvant. Examples of such curves can be found in Wark and Cox [WAR 35].

6.1.3. *Criteria for the occurrence of flotation*

There are two criteria for flotation to take place: an energetic criterion and a hydrostatic criterion.

1) energetic criterion: looking at the discussion in section 2.2.5 of [DUR 15c] of this series, we see that if the energy of adherence β of the water to the ore particle is negative, the water spontaneously separates from the ore. However, as we shall see in section 6.1.5, generally this is only possible after the addition of a "collector" to the pulp, with the collector affixing to the surface of the grains of ore and rendering them hydrophobic.

2) hydrostatic criterion: let us set:

v_p and v_B: volumes of a solid particle and of a bubble (m^3)

ρ_p and ρ_L: densities of the particle and of the liquid (kg.m^{-3})

n_p: number of particles affixed to a bubble.

Let us write that the Archimedes thrust (buoyancy) exerted on the ensemble of bubbles and particles is greater than the weight of the particles. After division by the acceleration due to gravity g, we obtain:

$$\left(n_p v_p + v_B\right)\rho_L > n_p v_p \rho_p$$

Thus:

$$v_p < \frac{v_B}{n_p}\left(\frac{\rho_L}{\rho_p - \rho_L}\right)$$

If we bring into play the diameters d_p of the particle and d_B of the bubble, we obtain:

$$d_p < d_B \left[\frac{\rho_L}{n_p\left(\rho_p - \rho_L\right)}\right]^{1/3}$$

For example, if the quantity in square brackets here is equal to 1/30, we should have:

$$d_p < 0.32 d_B$$

This explains why, for bubbles whose diameter is between 0.5mm and 5mm, the diameter of the particles is:

Metallic ores $0.2 \text{ mm} < d_p < 0.3 \text{ mm}$

Non-metallic ores $0.6 \text{ mm} < d_p < 1.6 \text{ mm}$

Indeed, metallic ores are generally denser than non-metallic particles such as sulfur or graphite.

6.1.4. *Dispersion of air in the cell*

The flotation cell is mechanically aerated. An energetic stirrer disperses the air introduced beneath the thruster in the form of small bubbles. The pulp is introduced beneath the froth and is underdrawn at the base of the cell. If we wish to install several cells in a series, the pulp can be pumped from one cell to the next under the influence of level regulation, or else by gravity if there is a significant difference in level between two consecutive cells. However, in the Denver cell, the pulp is fed in and removed almost at exactly the same level (at the bottom), but the thruster of the stirrer, in addition to the dispersion of the air, creates a motion of internal circulation in the cell through a draft tube which is coaxial with the axis of the thruster.

For the dispersion of air, we use the following characteristic numbers:

Reynolds number:

$$Re = \rho_L \frac{ND^2}{\mu_L} \qquad \text{between } 10^6 \text{ and } 7 \times 10^6$$

Froude number:

$$Fr = \frac{N^2 D}{g} \qquad \text{between } 0.1 \text{ and } 5$$

Power number:

$$N_p = \frac{P}{\rho_L N^3 D^5} \qquad \text{between } 0.5 \text{ and } 5$$

Aeration number:

$$N_a = \frac{Q_G}{ND^3} \qquad \text{between } 0.01 \text{ and } 2$$

ρ_L: density of the liquid (kg.m^{-3})

N: rotation frequency of the thruster (rev.s^{-1})

D: diameter of the thruster

μ_L: viscosity of the liquid phase (Pa.s)

g: acceleration due to gravity: $9.81 m.s^{-2}$

P: power at the shaft of the stirrer (Watt)

Q_G: air flowrate ($m^3.s^{-1}$)

The coalescence of the bubbles in the pulp is a hampering phenomenon which decreases the surface area of bubbles available to the solid. However, coalescence only occurs if the surface tension of the liquid is high, but it is typical to add frothing agents to the pulp, which prevent such coalescence by reducing the surface tension of the liquid.

6.1.5. Collectors

The collector is the most important adjuvant. It adsorbs to the surface of each particle, making that particle hydrophobic. Xanthates and thiophosphates are the most common collectors, and they are used to float sulfides and phosphates. Alkyl sulfonates float oxides. Fuel oil or creosote are useful for sulfur, graphite and charcoal dust. Finally, amines float silica, mica, feldspar, etc. The quantity of collector added varies, from case to case, between 50 g and 500 g per tonne of ore.

The energy E_{ad} characterizing the adherence of the water to the solid is the sum of:

E_d: Van der Waals dispersion forces

E_h: hydrogen bonds on the polar sites of the solid

E_i: interaction with the ionic sites

When the ore is fractured, E_h and E_i become significant, and the role of collectors is to mask the corresponding sites, because these sites are hydrophilic. Thus, the molecules of collectors contain:

– a polar group which binds to the surface of the ore;

– a hydrophobic hydrocarbon chain.

Oxides and silicates have a significant (E_h + E_i) and are hydrophilic, whereas the (E_h + E_i) of sulfides is much lesser, so the latter float more easily. The result is that the carbonate chain of sulfide collectors is much shorter than that of silicate- and oxide collectors.

Certain ores (graphite, sulfur, talc, molybdenite, etc.) do not develop (E_h + E_i) after milling, and can be floated without a collector. We also say that, at the surface, all of their bonds are saturated.

6.1.6. *Other adjuvants*

Frothing agents lower the surface tension of the liquid phase, and thus favor the dispersion of the air bubbles, and stabilize the froth at the surface of the pulp. The most commonly used frothing agents are ethers of polypropylene glycol, cresylic acids, pine oil, and certain aliphatic alcohols. The amounts used vary between 5 g and 100 g per tonne of ore.

A *depressor*, unlike a collector, decreases the flotability of an ore. It is helpful when we wish to separate two ores. Thus, sodium cyanide "depresses" pyrite during the flotation of lead sulfide, zinc sulfide or copper sulfide. Lime depresses pyrite, and sodium silicate depresses quartz.

Regulators enable us to set the value of the pH for which flotability is optimal. They are either acids (H_2SO_4, HCl) or alkalis (soda, lime, Na_2CO_3 and sodium phosphates are some examples).

We can also mention *activators*, which facilitate the action of the collector by adsorbing to the surface of the particles. Copper sulfate, sodium sulfide and lead nitrate are activators.

All adjuvants are introduced into the pulp by prior mixing in a stirred vat, called a conditioner. Indeed, the idea is to provide sufficient contact time between the adjuvants and the ore that we wish to "float" (or more accurately, "make float").

The *impurities* present in the crude ore and therefore in the pulp are unintended adjuvants, which can significantly alter the operation of flotation (particularly the ascension rate of the bubbles).

6.1.7. *Flowrate of solid brought to the froth from the pulp*

When a bubble whose diameter is d_{BP} rises the height dh, it has swept a cylindrical volume:

$$\frac{\pi d_{BP}^2 dh}{4}$$

Let c_P represent the concentration of particles in the pulp (measured in m^{-3}). Given the agitation which predominates in the aeration zone, we consider this concentration to be the same throughout. The number of particles found, therefore, is:

$$\frac{\pi d_{BP}^2 c_P dh}{4}$$

Now let η_a represent the fixation yield – i.e. the bubbles affixed as a proportion of those found. The number of particles fixed to the bubble over the height dh is:

$$\frac{\pi d_{BP}^2 c_P \eta_a dh}{4}$$

η_a is probably of the order of 0.2 (or even 0.1).

However, fixation can only occur on the fraction $(1 - \chi)$ of the surface of bubbles not yet covered by particles. Let us accept the hypothesis, as does Cailloce [GAI 75], that the surface of contact of a particle with the bubble to which it is affixed is d_p^2 (where d_p is the diameter of a particle). Also let n_p be the number of particles affixed to a bubble. The fraction of the surface of the bubble covered by particles is:

$$\chi = \frac{n_p d_p^2}{\pi d_{BP}^2} = kn_p \qquad (k \ll 1)$$

Finally, the number of particles affixed over the height dh is:

$$dn_p = \frac{\pi d_{BP}^2 \eta_a c_P (1 - kn_p) dh}{4}$$

Let us integrate over the height H of the aeration volume:

$$n_p = \frac{1}{k}\left[1 - \exp\left(-\frac{d_p^2}{4}\eta_a c_p H\right)\right] = \frac{\pi d_{BP}^2 \chi}{d_p^2}$$ [6.1]

This gives us the fraction covered:

$$\chi = 1 - \exp\left(-\frac{d_p^2}{4}\eta_a c_p H\right)$$ [6.2]

The flowrate of bubbles in terms of number is:

$$\frac{6 Q_G}{\pi d_{BP}^3}$$

Q_G: gas flowrate ($m^3.s^{-1}$)

The flowrate in terms of number of particles brought to the froth from the pulp is:

$$q_P = \frac{6 Q_G}{\pi d_{BP}^3} n_p$$ [6.3]

6.1.8. *Flowrate in terms of number of particles in the froth*

In line with Cailloce [CAI 75], we shall agree that:

– the surface area of contact of a particle with the bubble to which it is affixed is d_p^2, as before:

d_p: diameter of a particle (m)

– in the froth, the surface of the bubbles is entirely covered by particles, whereas in the pulp, only the fraction χ of the surface of the bubbles is covered.

The bubble surface flowrate in the froth is:

$$S_M = \frac{6Q_M}{d_{BM}} = \frac{6(Q_G - Q_F)}{Cd_{BP}} \qquad (m^2.s^{-1}).$$

Q_F: flowrate of gas leakage above the froth.

Let us set:

$$\frac{d_{BM}}{d_{BP}} = C$$

d_{BM}: diameter of the bubbles in the froth (m)

The flowrate in number of particles in the froth is:

$$q_M = \frac{S_M}{d_p^2}\frac{6(Q_G - Q_F)}{Cd_B d_p^2} \qquad [6.4]$$

6.1.9. *Diameter of bubbles in the froth*

Some authors have attempted to take account of a certain degree of reflux of the particles from the froth into the pulp. Cailloce [CAI 75] even gives an expression for that reflux, but to do so, he is forced to introduce two empirical parameters. Here, the reflux thus expressed has not been taken into account, because its existence is tantamount simply to a decrease in the fixation yield η_a, which is a purely empirical parameter.

Thus, we can assume the equality of the flowrates in terms of number q_P and q_M given by equations [6.3] and [6.4]. Consequently:

$$\frac{d_{BM}}{d_{BP}} = C = \left(1 - \frac{Q_F}{Q_G}\right)\left[1 - \exp\left(-\frac{d_p^2}{4}\eta_a c_p H\right)\right]^{-1}$$

6.1.10. *Concentration of the pulp: material balance*

In light of equations [6.2] and [6.3], the flowrate of particles brought to the froth from the pulp is:

$$q_P = \frac{6Q_G}{d_{BP} \times d_p^2}\left[1 - \exp\left(-\frac{d_p^2}{4}\eta_a c_p H\right)\right]$$

However, if we discount the liquid brought by the froth, we also have:

$$q_P = Q_F c_F - Q_F c_p$$

Q_F: flowrate of pulp fed into the cell ($m^3.s^{-1}$)

c_F: concentration of particles in the feed (m^{-3})

c_p: concentration of particles in the cell pulp (m^{-3})

By making the two values of q_p equal to one another, we obtain an equation that can be used to calculate the concentration c_p – i.e. the number of particles per m^3 of pulp in the cell.

6.1.11. *Flotation parameters*

In view of the above, there are two such parameters:

– the fixation yield η_a;
– the gas leak flowrate Q_F.

6.1.12. *Battery of cells*

The cells are always laid out in a series on the side of the pulp. The number of cells needed is, obviously:

$$N = \frac{W_{mi}\tau\omega_p}{\Omega_f}$$

W_{mi}: flowrate of ore (ton.h^{-1})

τ: length of stay of the pulp in the flotation volume of the cell (h)

ω_p: volume of pulp per ton of ore (m^3.tonne^{-1})

Ω_f: volume devoted to flotation in a cell (m^3)

Flotation can take place in two steps:

– rough grinding, when a portion of the waste material is eliminated;

– finishing or cleaning, when we recover the ore, freed of the waste material.

In addition, though, we can envisage recycling the gangue material recovered during finishing.

Naturally, thickeners and/or filters are needed to recover the ore.

6.1.13. *Uses of flotation*

Flotation is used to concentrate material from sources poor in carbon or metal derivatives. For example, the process is used to concentrate:

– native metals or metalloids (copper, silver, gold), sulfides (iron, zinc, lead, copper, silver), nonsulfide minerals (cerussite, hematite, magnetite);

– charcoals which are separated from schist;

– particles which are too fine to be treated by cleansing operations.

We can also separate calcium phosphate from hydrophilic silica in this way.

Flotation can be used to enrich bituminous sand, degrease wool washwater, purify fruit juice, separate ink from the pulp of old papers and, generally, purify certain industrial liquors.

6.1.14. *Additional indications*

Bushell [BUS 62] puts forward several empirical equations to express the variation of the concentration of solid particles in the pulp.

Loveday [LOV 66] proposes a theoretical simulation of the whole of flotation. He brings gamma functions (see [SPI 92]) into play, and the Laplace transform (see [DEM 02]). Unfortunately, the author does not specify the diameter of the particles he is considering.

Wookburn *et al.* [WOO 71] studies the influence of the particle size distribution on flotation.

King *et al.* [KIN 74] proffer a model for the fixation of solid to bubbles.

Trahar and Warren [TRA 76] and King [KIN 75] examine the influence of particle size on their fixation to bubbles. Dobby and Finch [DOB 87] offer a complete theory for the phenomenon.

6.1.15. *Particle diameter*

Let us more generally examine the influence of the particle diameter.

Hydrophilic fine particles (gangue and silicates, in particular) are *entrained* by the water separating the bubbles, which decreases the effectiveness of separation.

Fine particles easily adsorb adjuvants, owing to their large surface area in relation to their mass. This is to the detriment of adsorption on coarser grains. It is only possible to satisfactorily collect coarse grains by increasing the concentration of the collector.

The quality of the flotation is the product of two probabilities:

– the probability of collision p_c, which increases with d_p;

– the probability p_s, which defines the solidity of adherence of the particles to a bubble, and which decreases if d_p increases.

Consequently, the flotation yield passes through a maximum as a function of d_p. In general, the quality of flotation can be defined thus:

$d_p < 10 \ \mu m$ difficult

$10 \ \mu m < d_p < 70 \ \mu m$ effective

$d_p > 70 \ \mu m$ difficult

Using numerous examples, Trahar [TRA 81] illustrates this influence of the particle diameter on flotation.

An anonymous author [ANO 86] proposes to intercalate a flotation operation between two rounds of milling, which, in certain cases, can improve the recovery of precious metals. More generally, McIvor and Finch [MCL 91] study the combination of flotation and milling.

6.2. Froth columns

6.2.1. *Operation*

The device commonly known as a froth column has a solution inlet at the top of the column, and an air inlet through a perforated plate at the bottom. The contact between the bubbles and the liquid takes place in a countercurrent direction, and the exhausted liquid is underdrawn at the base of the column.

The air is brought into the column by a compressor, through a perforated plate, or else a plate with nozzle tips of small diameter. With this latter arrangement, Kumar and Kuloor [KUM 67] propose the following for the volume of the bubbles:

$$\upsilon_B = k \frac{Q_G^{6/5}}{g^{3/5}} \qquad\qquad (m^3)$$

Q_G: gas flowrate $(m^3.s^{-1})$

g: acceleration due to gravity $(m.s^{-2})$:

$0.80 < k < 1.72$

Whereas flotation consists of attaching particles of ore to bubbles, in a froth column we extract certain solutes from a solution by exploiting the solute's property of surfactancy – i.e. its tendency to concentrate at the gas–liquid interface (it is also possible to extract non-surfactant components as long as they are combined, previously, with a surfactant adjuvant).

Note that the entrainment of the solute in the froth is comparable to stripping by an inert gas.

The equilibrium between the surface concentration of the surfactant solute and its concentration inside the liquid can generally be expressed by a Langmuirian isotherm.

$$\Gamma = \frac{ac}{1+bc}$$

Γ: surface concentration (kmol.m^{-2})

c: concentration in liquid (kmol.m^{-3})

By writing that the total solute in the froth is the sum of the solute at the surface of the films and the solute inside them, we obtain:

$$Q_{LM}c_{LM} = G\ S\ \Gamma + Q_L c_w$$

Q_{LM}: liquid flowrate of the settled froth (m^3.s^{-1})

Q_L: underflow of liquid at the bottom of the column (m^3.s^{-1})

c_{LM}: concentration of the settled froth (kmol.m^{-3})

G: gaseous flowrate (m^3.s^{-1})

S: surface of gas–liquid interfacial films in the froth per m^3 of froth (m^2.m^{-3})

c_W: concentration of solute at the bottom of the column (kmol.m^{-3}.)

Indeed, the interfaces are at equilibrium with the liquid at the bottom of the column, whose concentration is c_W.

The size of the bubbles is between 1 and a few mm. The volumetric fraction of liquid in the froth must be less than 10%. The size of the bubbles is evaluated by observation immediately beneath the froth.

The coalescence of the bubbles in the froth is a hindering phenomenon about which not much is known. In the froth, even slight stresses can cause the rupture of the films, as can their thinning by the drainage of the liquid. In addition, the gas diffuses from the small bubbles to the larger ones through the liquid films, because in accordance with Laplace's law, the pressure is higher when the bubble is small.

It is possible to reduce coalescence by using a small height of froth and a high gaseous flowrate, to reduce the length of stay of the froth and, consequently, the drainage of the liquid.

Lemlich's article [LEM 68] offers an example of the calculation of the air flowrate, the surfactant flowrate and the diameter of the column to reduce the concentration of the feed tenfold.

Fanlo et al. [FAN 65] give a relation that can be used to calculate the material transfer coefficient between the rising liquid and the falling liquid in the hypothetical case of a reflux (which we shall not discuss here).

Expulsion of a Liquid by Pressing

7.1. Mechanism for pressing inorganic DSs

7.1.1. *An ideal representation: unidirectional pressing*

A bed of particles (or plant matter) is sandwiched between a pressing piston and a rigid, perforated filter medium.

F_{pr}: pressing force (N)

F_{fil}: force on the filter medium (N)

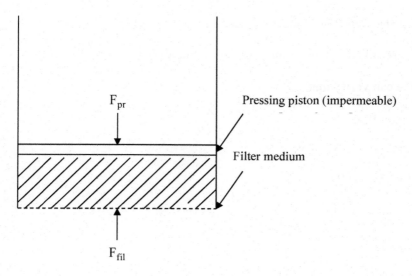

Figure 7.1. *Unidirectional pressing*

With the exception of the case discussed in section 7.1.2, we assume that the divided solid is *completely saturated* with liquid – i.e. that it contains no pockets of gas.

7.1.2. *Law for pressing solids*

We first assume the DS is in its dry state – in other words, that no liquid is present. The solid matrix, when subjected to pressure, will experience a decrease in apparent volume, where:

P_s: pressure exerted on the solid above atmospheric pressure: bar G

V_0: volume of the DS for Ps = 0 (m³)

V_m: minimum volume of the solid matrix (for infinite Ps) (m³)

The law usually applied for the variation in the volume V based on the pressure P is [NAG 79, NAG 86]:

$$\frac{V_0 - V_m}{V - V_m} = Ln\left(\frac{P_s}{P^*} + e\right)$$

[7.1]

e is the Neperian logarithm base.

P* is linked to the compressibility of the dry porous solid.

We can verify that:

If $P_s = 0$ $V = V_0$

If $P_s \rightarrow \infty$ $V \rightarrow V_m$

We now set:

$$\pi = Ln\left(\frac{P_s}{P^*} + e\right); \quad \alpha = \frac{V_m}{V_0} = \frac{1 - \varepsilon_0}{1 - \varepsilon_m}; \quad \frac{V}{V_0} = \frac{1 - \varepsilon_0}{1 - \varepsilon}$$

Equation [7.1] becomes:

$$1 - \varepsilon = \frac{\pi(1 - \varepsilon_0)}{1 + \alpha\pi}, \text{ meaning that } \varepsilon = f(P)$$

7.1.3. *Pressure of fluids*

We assume that the plant matter is compressed between a flat plate and a perforated grid. The solid is partially saturated with liquid.

The solid is assumed to have an identical relative variation of apparent volume throughout its entire thickness. In other words, it is uniformly pressed across this thickness.

The position of a solid slice is characterized by its distance h from the plate. During compression, the following ratio remains *invariable*:

h/Z

Z: instantaneous thickness of the plant bed

The total volume (gas + liquid) of fluids passing through the slice located at the instantaneous depth h is, per unit time and area of the plate:

$$Q_{g\ell} = \frac{h}{Z}\frac{dZ}{d\tau} \quad \text{where} \quad \frac{dZ}{d\tau} < 0$$

According to Darcy's law, if this overall throughput were entirely liquid, the friction through the elementary slice of thickness dh would cause the pressure to drop:

$$dP_L = \frac{-\mu}{K}Q_{g\ell}dh = \frac{\mu}{K}\frac{h}{Z}\left(\frac{dZ}{d\tau}\right)dh$$

For the overall two-phase flow, using Lockhart and Martinelli's [LOC 49] formula, for example, it is necessary to modify dP_L by a coefficient ϕ^2.

The overall pressure of the fluids is zero at the grid where h = Z and maximum at the plate where h = 0. It increases linearly between the two.

The overall pressure at the plate is obtained by integration:

$$P_{gl} = \frac{\phi^2\mu}{KZ}\left(\frac{dZ}{d\tau}\right)\int_Z^0 h\,dh = -\frac{\phi^2\mu}{K}\left(\frac{dZ}{d\tau}\right)\frac{Z}{2} > 0$$

In reality, the coefficient ϕ^2 depends on the density of the gas, which, itself, depends on the pressure. The permeability K depends on Z. We must therefore carry out a numerical integration.

7.1.4. *Theoretical process of pressing*

We shall use the following parameters:

A: surface area of the plate (m^2)

ε: porosity of the solid

P_s: compression stress on the solid

The force exerted on the plate is:

$$F_{pl} = A(1-\varepsilon)P_s + A\varepsilon P_{g\ell}$$

As the fluids are evacuated through the grid, we see that the porosity decreases, so that:

P_s increases and P_{gl} also increases

If the force F_{pl} is imposed and is constant, at the start of the process, the discharge the fluids will offer most resistance but this will decrease the porosity, and the compression of the solid will take over.

If the movement of the plate over time is imposed, then ε, P_s and P_{gl} can be deduced. The pressures of both P_s and P_{gl} increase up to the limit imposed by the mechanical strength of the press.

For inorganic products, it is possible to use the reasoning employed in sections 7.1.2, 7.1.3 and 7.1.5.

However, we shall see that for plants, the process of pressing is substantially differently to this theoretical one (see section 7.2).

NOTE.–

Shirato *et al.* [SHI 79, SHI 86] propose semi-empirical solutions for the result of pressing.

7.1.5. *Equipment used for pressing inorganic dispersions*

A non-exhaustive list of possible equipment would include:

1) The cake-compression filter press:

At the end of the filtration – i.e. when the cake fills the entire chamber – the volume of the chamber is reduced, compressing the cake. For this purpose, one of the chamber walls is an elastomer membrane behind which water or air pressure of 15 to 20 bar is applied.

2) The tubular press:

This works on the same principle as above. The cake occupies the annular space between two cylindrical walls. The inner wall is made of elastomer and can withstand pressures of 200 bars.

3) The band press:

The product is sandwiched between two permeable conveyor belts moving at the same speed. There are two possible devices.

If the movement of the bands is non-continuous, every time the device stops, both the bands and the product are compressed between two plates.

If the movement is continuous, the pressing force is obtained from pairs of rollers whose surfaces are progressively closer together from one pair to the next.

7.2. Pressing of plant matter

7.2.1. *Advantage of pressing*

This operation is used to extract an active ingredient from a plant, when that ingredient is in liquid form. It might be:

– an aqueous juice;

– an oil.

Pressing requires much less energy than evaporation. It takes 6 kJ to raise 1 kg of water to the pressure of 70 bars, while it takes almost 3000 kJ to evaporate it.

7.2.2. *Terminology*

An oilseed, for our purposes, means a nut that essentially comprises a hard shell (of greater or lesser thickness), within which there is a kernel containing the oil. These kernels are ground, adequately moistened and subjected to pressing, which provides:

– the crude oil with suspended water and solids (sludge or "feet");

– the de-oiled solid residue called filterpress cake, which includes shell debris if the shell has not previously been removed.

Olive oil is the exception to the above because it comes from the pulp of the fruit, rather than the kernel. The solid residue from the pressing is known as "olive cake", and includes debris from the shell.

The palm fruit is treated in much the same way as olives – i.e. it is crushed and pressed – but the solid residue from the fruit pulp is called duff. The mat contains shell debris, and the whole forms the filterpress cake.

When a fruit is pressed to extract the juice (apple, grape), the solid residue is called the must.

7.2.3. *Mechanism of pressing*

Pressing of organic matter involves three steps:

1) expulsion of the interstitial gas;

2) compression of the solid matrix;

3) expulsion of the juice.

Figure 7.2 illustrates this process.

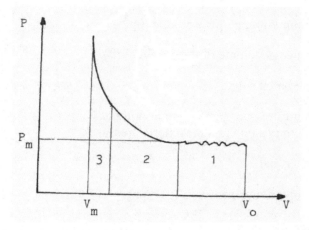

Figure 7.2. *Compression of plant matter*

This figure shows the variations in the pressure applied to the product – by a piston, for example – depending on the product's volume:

– First step:

This step takes place at constant pressure P_m. The product contains some gas pockets at low pressure. When the wall of a pocket gives way, solid and liquid will enter, decreasing the volume of the product. The fact that the pockets burst in isolation explains the serrated shape of the curve for step 1. No juice appears during this step.

– Second step:

The remaining gas is compressed or continues to escape but, here, the solid matrix offers compression resistance according to the empirical law (equation [7.1]):

$$\frac{V_0 - V_m}{V - V_m} = Ln\left[\frac{P - P_0}{P^*} + e\right]$$

V_0: initial volume of the product (m³)

V_m: minimum volume of the product (m³)

V: instantaneous volume of the product (m³)

P: pressure: bar absolute

P_0: minimum pressure below which nothing happens: absolute bar (P_0 is slightly higher than ambient atmospheric pressure)

P^*: parameter: bar

e: natural logarithm base

– Third step:

The solid matrix continues to be compressed but there is now discharge of the liquid, whether by destruction of an increasing number of cells or by exudation through the walls of the cells remaining intact. Cell destruction leads to an increase in open porosity and a consequential decrease in closed porosity. The sum of these two porosities – that is to say, the total porosity – varies little, but the increase in open porosity releases the fluid and encourages it to flow through the product. Finally, a significant increase in pressure corresponds to a moderate decrease in product volume.

The cell debris entrained by the juice increases in quantity with:

– pressure which, as we have just seen, destroys the cells;

– temperature, which kills the cells and weakens their walls.

If the movement of the piston is halted, the pressure rapidly decreases to a value greater than P_0 because the fine particles are distributed differently between the granules or, more accurately, the cell debris are rearranged between the cells. This is called relaxation.

7.3. Equipment used for pressing plants

7.3.1. *Hydraulic press*

The flour resulting from the milling of oily kernels is molded into parallelepipeds whose horizontal surface is a square and whose thickness is greater than 2.5 cm.

This molded flour is then wrapped in a press cloth and placed between two plates. Presses may contain 15 stacked plates, guided by four vertical columns. They can process 300 kg of flour per operation at a rate of approximately two operations per hour.

The advantage of these machines is that the compression is gradual, because two fluid motors are used: one at low pressure and one at high pressure. The material releases its oil gradually and contains only little "foot". The press cloths do not wear out prematurely.

In hydraulic presses, the flour can be subjected to a final pressure of about 300 bar. The yield obtained is therefore less than that of screw presses where the pressure is higher.

7.3.2. *Continuous screw press*

This machine includes a cylindrical cage (the sheath) on a horizontal axis in which a screw pushes the product forward. This requires sufficient friction between the product and the wall of the cage to prevent the product turning with the screw instead of advancing. The diameter of the shaft increases in the direction of the product's progression, which reduces the volume available to it and creates compression. The compression ratio may be around 4 (initial volume over final volume).

If seeds are pressed to extract oil, the sheath must be able to withstand pressures of about 2000 bar. It therefore consists of bars separated by slots.

However, for fruit pressing, a thick perforated metal is sufficient because the pressure does not exceed 100 bar.

In the treatment of seeds, the machine must be cooled, so water circulates through the hollow shaft. The cage itself is strengthened by rings that are also cooled, and this cooling is passed on to the bars by conduction. It is necessary to evacuate the heat. because the considerable friction causes heating which may cause the oil and cake to darken, and could even cause the latter to combust. The cake comes out at a temperature varying between 95°C and 150°C, so it must be cooled with a water jet.

In order for the throughput of the machine to be satisfactory, the feed must be as dense as possible. Thus, the cross-section of the chute must be equal to that of the feed hopper. The hopper must always be full, even under slight pressure, to prevent the product from backing up. At this point, the air present between the flour flakes is also expelled, and the drainage slots are wide to accommodate a significant flow of oil.

The sheath becomes worn and polished, which reduces the oil yield and the flowrate of product, which begins to slip and rotate with the shaft. It is therefore advantageous to clean the product to remove any grit. Wear of the sheath affects the throughput and the yield.

To prevent the product from turning when the sheath becomes worn, manufacturers offer presses with two parallel screws rotating in opposite directions, but this creates pull-outs which generate sludge in the oil obtained.

A high rotation speed increases the production rate but reduces the oil yield by reducing the residence time in the machine. The rotation speed varies between machines from 20 to 50 rev.mn^{-1}.

The cake should be as thin as possible to facilitate drainage. This thickness therefore should not depend on the size of the machine so that the flowrate of the latter is proportional to the diameter of the screw. Screw presses can handle 2 to 20 tons.h^{-1} of product and consume between 10 and 20 kW per ton.h^{-1} of product treated.

To maintain the pressure in the machine, a cone is pushed into the outlet orifice to varying extents. This cone is well clamped. which increases the pressure and thus the yield, but may cause significant breakage of the shells if they are present, which increases internal wear. Electrical consumption obviously increases with the clamping of the cone. However, the electricity needed for normal operation of a screw press is half that which is needed for a load-based hydraulic press.

The radial pressure gradient is what causes the drainage of the oil. At a given distance along the axis, the pressure includes a permanent component and a periodic component which fluctuates with the passage of the screw threads. The permanent component is proportional to the axial pressure and varies along the axis as shown in Figure 7.2.

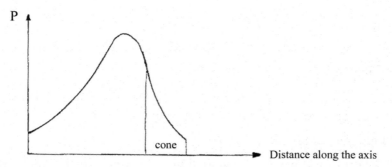

Figure 7.3. *Variations in pressure along the axis*

The radial pressure decreases with:

– the opening of the slots in the cage;

– the wearing of the screw.

This pressure has the following values:

– for pre-pressing before solvent-based extraction: 250 to 300 bar (which leaves 15 to 20% of oil in the cake);

– for pressing to achieve full extraction: 400 to 2000 bar (which leaves less than 5% of oil in the cake).

In general, seeds are processed by pressing if their oil content exceeds 20%.

7.3.3. *Other continuous presses*

1) Cylinder presses for sugarcane:

These machines, called bagasse mills, crush the plant between two cylinders. The force bringing the cylinders together is:

$$F = \lambda \, L \, D$$

L and D are the length and diameter of the cylinders.

If we assume that the force is distributed over a band whose width is equal to one tenth the diameter of the rollers, we can define a specific pressure:

$$P_s = \frac{F}{0.1 \times L \times D} = 10\lambda$$

This pressure is of the order of 200 bar.

The roller diameter varies from approximately 50 cm to 1m, and a pair of rollers can handle 50 to 500 t.h^{-1} of sugarcane.

2) Pulp presses for paper:

The sheet of wet paper is supported by a felt band which passes between the two cylinders along with it. The felt draws the liquid out of the sheet and passes it to the cylinder, which is perforated.

The pressure reaches a maximum in front of the plane of the axes of the rollers. After the plane, the paper and felt decompress. There is then negative pressure in the sheet and, because of capillary pressure, there is a transfer of liquid from the felt to the sheet, whose pores are smaller. This is known as rewetting.

Rewetting also occurs in sugar mills. Although there is no felt, with the plant being crushed between the two cylinders, we see "squirting" or "extrusion" of juice downstream. The negative capillary pressure of the downstream decompressed product helps reabsorb the juice.

3) Belt presses:

The product, while being carried along a conveyor belt, is pressed successively by a series of pairs of rollers, with the pressure increasing as the space between them decreases.

In these machines, the pressure exerted is limited, so they are used for fruit juices and wastewater treatment sludge.

7.3.4. Characteristics of oilseeds

The less oil kernels contain, they harder and more compact they are. In addition:

1) tough fibrous seeds provide high-viscosity oil, as in the case of coconut oil and palm oil;

2) soft seeds provide a fluid oil, as is the case with sunflower, peanuts and especially rapeseed;

3) oil is also extracted from the pulp (rather than the seed) of certain fruits, and the product obtained is an oil–water emulsion; as in the case of olives and palm fruit. These oils are fluid.

In general, the optimal humidity of the flour before pressing increases with oil content.

The threads of a continuous press screw are spaced further apart for soft products than for hard ones. For the latter, a slightly rougher surface for the cage material is acceptable.

Fibrous products produce little sludge (unlike soft seeds such as peanuts).

7.3.5. Maximum pressure depending on product

There is a maximum pressure for each product, regardless of which type of machine is used as, when this pressure is attained, the porosity available for the flow of fluid becomes very small, as does the average pore radius, which greatly reduces the resulting liquid flow.

In addition, in the case of plants from which aqueous juice is extracted (essentially fruits), the initial moisture is considerable and it is often unnecessary to aim for higher pressures to extract the last few drops of juice because the production gain would not justify the investment.

Here are a few examples of maximum pressures:

– Oilseeds: 1000–2000 bar

– Lucerne: 100 bar

– Apple pulp: 30–45 bar

– Grape: 20 bar

7.3.6. *Pressing adjuvants*

The rigidity of the particles and their irregular form, which is far removed from a sphere, means that pressing adjuvants have large open porosity if they are mixed with a product under compression. Thus, there are still pores with a sufficient radius to allow the flow of liquid.

They are used for pressing apple juice when the fruit is very ripe, because the insoluble protopectin has, by that point, been transformed into soluble pectin, which makes the ground pulp soft and viscous, hindering the drainage of the juice.

The adjuvants used in the food industry are the following:

1) Cellulose: can be used either as 1mm-long fibers, or in the form of seed shell debris of varying dimensions. Fibers are used for pressing apples and shell debris in the pressing of some oilseeds.

2) Diatomite: is obtained by grinding the fossilized shells of marine animals.

3) Perlite: comes from the grinding of a volcanic rock with a glasslike structure which looks like glass debris.

4) Polyolefin fibers: are around 1 mm long, with a diameter of 10 to 15 μm. Unlike the other adjuvants mentioned, these fibers have the ability to absorb certain chemical species and have an area of 4 to 6 m^2 per gram. They replace asbestos, which is banned in food products.

7.3.7. *Factors influencing the yield when pressing oils*

To obtain a high yield, it is advantageous to have:

– a thin cake, which shortens the path traveled by the expelled oil;

– high final pressure to break up the cells as much as possible. However, we have seen that there is a limit to this pressure;

– adequate pressing time to allow the liquid time to flow without causing fine particles;

– an elevated temperature, which kills cells and facilitates exudation;

– sufficient humidity to soften the pectic cement by hydrolysis; but if the humidity exceeds a certain limit, the yield decreases;

– preliminary grinding to facilitate fluid transport into the parenchymal tissue; however, if grinding is taken too far, the fine particles can clog the pores of the product;

– a filtration adjuvant to facilitate the percolation of the oil; thus, we obtain a higher yield with unshelled soy beans.

Let us take a look at some representative yield values.

The juice yield, expressed in relation to the mass of treated product, has the following approximate values:

– Destemmed grape: 0.75

– Apple: 0.65

– Tomato: 0.70

– Citrus fruits: 0.40

– Pineapple: 0.50

APPENDICES

Appendix 1

Numerical Integration
4th-order Runge–Kutta Method

With the 4th-order Runge–Kutta method, we integrate the differential equation:

$$\frac{dx}{d\tau} = F(x, \tau)$$

$$x_{\tau=0} = x_0$$

We start by setting:

$$\tau_{i+1/2} = \tau_i + \frac{\Delta\tau}{2}$$

$$x_{i+1/2}^{(1)} = x_i + \frac{\Delta\tau}{2} F(x_i, \tau_i)$$

$$x_{i+1/2}^{(2)} = x_i + \frac{\Delta\tau}{2} F(x_{i+1/2}^{(1)}, \tau_{i+1/2})$$

$$x_{i+1}^{(1)} = x_i + \Delta\tau \, F(x_{i+1/2}^{(2)}, \tau_{i+1/2})$$

Hence:

$$x_{i+1} = x_i + \frac{\Delta\tau}{6}\Big[F\,(x_i, \tau_i) + 2F\,(x_{i+1/2}^{(1)}, \tau_{i+1/2})$$

$$+2F\,(x_{i+1/2}^{(2)}, \tau_{i+1/2}) + F\,(x_{i+1}^{(1)}, \tau_{i+1})\Big]$$

We can generalize the method to apply to a system of n 1st-order differential equations involving n variables x_j (j ranging between 1 and n). The independent variable is x_0.

$$\frac{dx_j}{dx_0} = F_j\left(x_0, x_1, ..., x_j, ..., x_n\right)$$

Let us set:

$$x_{0,i+1} = x_{0,i} + \Delta x_0 \quad \text{and} \quad x_{0,i+1/2} = x_{0,i} + \frac{\Delta x_0}{2}$$

$$x_{j,i+1/2}^{(1)} = x_{j,i} + \frac{\Delta x_0}{2} F_j\left(x_{0,1}, ..., x_{j,i}, ...x_{n,i}\right)$$

$$x_{j,i+1/2}^{(2)} = x_{j,i} + \frac{\Delta x_0}{2} F_j\left(x_{0,i+1/2}, ..., x_{j,i+1/2}^{(1)}, ..., x_{n,i+1/2}^{(1)}\right)$$

$$x_{j,i+1}^{(1)} = x_{j,i} + \Delta x_0 F\left(x_{0,i+1/2}, ..., x_{j,i+1/2}^{(2)}, ..., x_{n,i+1/2}^{(2)}\right)$$

and finally:

$$x_{j,i+1} = x_{j,i} + \frac{\Delta x_0}{6}\Big[F_j\left(x_{0,i}, ...x_{j,i}, ..., x_{n,i}\right) + 2F_j\left(x_{0,i+1/2}, ...x_{j,1+1/2}^{(1)}, ...x_{n,i+1/2}^{(1)}\right)$$

$$+2F_j\left(x_{0,i+1/2}, ...x_{j,i+1/2}^{(2)}, ...x_{n,i+1/2}^{(2)}\right) + F_j\left(x_{0,i+1}, ...x_{j,i+1}^{(1)}, ...x_{n,i+1}^{(1)}\right)\Big]$$

Appendix 2

The CGS Electromagnetic System

A2.1. Potential in the international system (SI) and in the CGS EM system

By definition (if $1C = 1$ Coulomb)

$$1\underset{\text{SI}}{C} = 3.10^9 q_{\text{CGSEM}} \text{ and } \varepsilon_0 = \frac{1}{36\pi.10^9} C.m^{-1}.V^{-1}$$

The potential created in a vacuum by a 1 C charge at a distance of 1 m is:

$$\text{Pot}_{\text{SI}} = \frac{1C}{4\pi\varepsilon_0 1m} = 9.10^9 \text{ Volt}$$

The potential created by a charge q of 1 CGSEM unit at the distance of 1 cm is:

$$\text{Pot}_{\text{EMCGS}} = \frac{q}{1cm} = 1 \text{ volt stat}$$

The ratio between these two potentials must be such that:

$$\frac{\text{Pot}_{\text{SI}}}{\text{Pot}_{\text{CGSEM}}} = \frac{C}{q} \times \frac{cm}{m}$$

This means that:

$$\frac{9.10^9 \, \text{Volt}}{1 \, \text{volt stat}} = \frac{C}{q} \frac{cm}{m} = \frac{3.10^9}{100}$$

Hence:

$$1V = \frac{1 \, \text{volt stat}}{300}$$

NOTE.–

We can very simply write:

Volt × C = Joule

Volt stat × q = erg

Thus:

$$\frac{1 \, \text{Volt}}{1 \, \text{volt stat}} = \frac{\text{Joule}}{\text{erg}} \times \frac{q}{C} = 10^7 \times \frac{1}{3.10^9} = \frac{1}{300}$$

A2.2. Other units in SI and CGSEM

1) The conductivity κ is measured in $\Omega^{-1}.m^{-1}$. The conductance is the quotient of the intensity by the potential.

Finally:

$$[\kappa] = \frac{I}{VL} = \frac{C}{V.s.m} \qquad \text{(SI)}$$

However:

$$1 \, C = 3.10^9 \, q_{CGSEM}$$

$$1 \, V = \frac{1}{300} \, \text{volt stat}$$

$$1 \, m = 100 \, cm$$

Therefore:

$$\kappa_{SI} = \frac{3.10^9 \times 300}{100} \kappa_{CGSEM} = 9.10^9 \kappa_{CGSEM}$$

2) Although viscosity is not an electromagnetic value, here we state the correspondence between the CGS system and the international system. Viscosity is measured in Pa.s, which is to say in $kg.m^{-1}.s^{-1}$.

However:

1 kg = 1000 g

1 m = 100 cm

Thus:

1 Pa.s $= 10 \ g.cm^{-1}.s^{-1} = 10$ poise $= 10$ barye.s

Appendix 3

Characteristics of Screening Surfaces

Mesh aperture (mm)	Wire diameter (mm)
100	20
80	16
70	12
60	12
50	11
45	10
40	8
35	6
30	6
25	5
20	4
15	4
12	4
8	2.5
5	2.2
4	2
3	1.6
2	1.25
1.5	1
1	0.8
0.75	0.6
0.6	0.5
0.5	0.4
0.4	0.35
0.3	0.25
0.2	0.15
0.15	0.12

Appendix 4

Definition and Aperture of Sieve Cloths

NF X 11-501 ET P 18-304		ASTM E 11-39 ET AFA		TYLER STANDARD SCREEN SCALE		BSA 410		DIN 1171	
Reference	Aperture	Reference	Aperture	Reference	Aperture	Reference	Aperture	Reference	Aperture
Module	mm	No.	Mm	Mesh	mm	No.	Mm	No.	Mm
17	0.040	3.5	5.66	2½	7.925	300	0.053	0.060	0.060
18	0.050	4	4.76	3	6.680	240	0.066	0.075	0.075
19	0.063	5	4	3½	5.613	200	0.076	0.090	0.090
20	0.08	6	3.36	4	4.699	170	0.089	0.100	0.100
21	0.100	7	2.83	5	3.962	150	0.104	0.120	0.120
22	0.125	8	2.38	6	3.327	120	0.124	0.150	0.150
23	0.160	10	2	7	2.794	100	0.152	0.200	0.200
24	0.200	12	1.68	8	2.362	85	0.178	0.250	0.250
25	0.250	14	1.41	9	1.981	72	0.211	0.300	0.300
26	0.315	16	1.19	10	1.651	60	0.251	0.400	0.400
27	0.400	18	1	12	1.397	52	0.295	0.430	0.430
28	0.500	20	0.84	14	1.168	44	0.353	0.500	0.500
29	0.63	25	0.71	16	0.991	36	0.422	0.600	0.600
30	0.80	30	0.59	20	0.833	30	0.500	0.750	0.750
31	1	35	0.50	24	0.701	25	0.599	1	1
32	1.25	40	0.42	28	0.589	22	0.699	1.200	1.200
33	1.60	45	0.35	32	0.495	18	0.853	1.500	1.500
34	2	50	0.297	35	0.417	16	1.003	2	2
35	2.50	60	0.250	42	0.350	14	1.204	2.500	2.500
36	3.15	70	0.210	48	0.295	12	1.405	3	3
37	4	80	0.177	60	0.246	10	1.676	4	4
38	5	100	0.149	65	0.208	8	2.057	5	5
		120	0.125	80	0.175	7	2.411	6	6
		140	0.105	100	0.147	6	2.812		
		170	0.088	115	0.124	5	3.353		
		200	0.074	0150	0.104				
		230	0.062	170	0.089				
		270	0.053	200	0.074				
		325	0.044	250	0.061				
		400	0.037	270	0.053				
				325	0.043				
				400	0.038				

Starting at the top of a system of superposed sieves, the apertures are in a geometric progression, with the ratio r being less than 1.

According to the European norms NF, DIN and BSA, the ratio is:

$$r \# \frac{1}{\sqrt[10]{10}} = 0.8$$

For the American norms ASTM and TYLER:

$$r \# \frac{1}{\sqrt[4]{2}} = 0.84$$

Bibliography

[ANO 86] ANONYMOUS, "Flash flotation", *International Mining*, 14–17 May 1986.

[APP 84] APPLEGATE L.E., "Membrane separation processes", *Chemical Engineering*, pp. 64–89, 11 June 1984.

[AUL 80] AULAS F., RUMEAU M., RENAUD M. *et al.*, "Application de l'ultrafiltration à la récupération de cations métalliques en solution", *Informations Chimie*, vols. 204–205, pp. 145–152, August–September 1980.

[BAB 80] BABCOCK W.C., BAKER R.W., LA CHAPELLE E.D. *et al.*, "The mechanism of uranium transport with a tertiary amine", *Journal of Membrane Science*, vol. 7, pp. 71–87, 1980.

[BAL 82] BALIAN R., *Du microscopique au macroscopique*, Ellipses, 1982.

[BAN 03] BANDINI S., VEZZANI D., "Nanofiltration modeling: the role of dielectric exclusion in membrane characterisation", *Chemical Engineering Science*, vol. 58, pp. 3303–3326, 2003.

[BAR 74] BARTLETT D.R., MULAR A.L., "Dependence of flotation rate on particle size and fractional mineral content", *International Journal of Mineral Processing*, vol. 1, pp. 277–286, 1974.

[BER 61] BERGSMA F., KRUISSINK CH.A., "Ion-exchange membranes", *Fortschritte Hochpolymer Forschung*, vol. 2, pp. 307–362, 1961.

[BOW 95] BOWEN W.R., JENNER F., "Electroviscous effects in charged capillaries", *Journal of Colloid and Interface Science*, vol. 173, p. 388, 1995.

[BOW 96] BOWEN W.R., WILLIAMS P.M., "The osmotic pressure of electrostatically stabilized colloidal dispersions", *Journal of Colloid and Interface Science*, vol. 184, pp. 214–250, 1996.

[BRU 89] BRUN J.P., *Procédés de séparation par membrane*, Masson, 1989.

[BUS 62] BUSHELL C.H.G., "Kinetics of flotation", *Transactions of the Society of Mining Engineers of AIME*, vol. 223, no. 3, p. 266, 1962.

[CAB 03] CABANE B., HÉNON S., *Liquides, solutions, dispersions, émulsions, gels*, Éditions Belin, 2003.

[CAI 75] CAILLOCE A., Etude de la cinétique de flottation. Influence de la mousse en surface. Drainage du liquide interstitiel, Doctoral Thesis, University of Nancy I, 1975.

[CAR 37] CARMAN P.C., "Fluid flow through granular beds", *Transactions of the Institution of Chemical Engineers*, vol. 15, pp. 150–165, 1937.

[CHA 13] CHAPMAN D.L., "A contribution to the theory of electrocapillarity", *Philosophical Magazine*, vol. 25, p. 475, 1913.

[CHE 55] CHEN C.Y., "Filtration of aerosols by fibrous media", *Chemical Reviews*, vol. 55, p. 595, 1955.

[CHR 85] CHRISTOFOROU C.C., WESTERMANN-CLARK G.B., ANDERSON J.L., "The streaming potential and inadequacies of the Helmholtz equation", *Journal of Colloid and Interface Science*, vol. 106, no. 1, p. 1, 1985.

[COE 16] COE H.S., CLEVENGER G.H., "Methods for determining the capacities of slime-settling tanks", *Transactions of the American Institute of Mining Engineers*, vol. 55, pp. 356–384, 1916.

[CUS 71] CUSSLER E.L., EVANS D.F., MATESICH M.A., "Theoretical and experimental basis for a specific countertransport system in membranes", *Science*, vol. 172, p. 377, 1971.

[CUS 84] CUSSLER E.L., *Diffusion Mass Transfer in Fluid Systems*, Cambridge University Press, 1984.

[DAR 02] DARNON E., BELLEVILLE M.P., RIOS G.M., "Modeling ultrafiltration of complex biological solutions", *AIChE Journal*, vol. 48, p. 1727, 2002.

[DAV 78] DAVIES L., DOLLMORE D., "Sedimentation of suspension: factors affecting the dispersion and hindered settling of calcium and other carbonates in a variety of liquids", *Powder Technology*, vol. 19, pp. 1–6, 1978.

[DEM 02] DEMENGEL G., *Transformations de Laplace*, Ellipses, 2002.

[DER 37] DERJAGUIN B.V., "A theory of interaction of particles in presence of electric double layers and the stability of lyophobe colloids and disperse systems", *Acta Physicochimica URSS*, vol. 10, no. 3, 1937.

[DER 41] DERJAGUIN B.V., LANDAU L.D., "Theory of strongly charged lyophobic sols and of the adhesion of strongly charged particle in solution of electrolytes", *Acta Physicochimica*, vol. 14, p. 633, 1941.

[DOB 87] DOBBY G.S., FINCH J.A., "Particle size dependence in flotation derived from a fundamental model of the capture process", *International Journal of Mineral Processing*, vol. 21, p. 241, 1987.

[DON 11] DONNAN F.G., "Theorie der Membran Gleichgewichte und Membran potentiale", *Zeitschrift für Elektrochemie*, vol. 17, p. 572, 1911.

[DRE 72] DRESNER L., "Some remarks on the integration of the extended Nernst–Plank equation in the hyperfiltration of multicomponent solutions", *Desalination*, vol. 10, p. 27, 1972.

[DUF 78] DUFFEY M.E., EVANS D.F., CUSSLER E.L., "Simultaneous diffusion of ions and ion pairs across liquid membranes", *Journal of Membrane Science*, vol. 3, p. 1, 1978.

[DUR 53] DURAND E., *Electrostatique et magnétostatique*, Masson, 1953.

[DUR 99] DUROUDIER J.-P., *Pratique de la filtration*, Hermès, 1999.

[DUR 16a] DUROUDIER J.P., *Thermodynamics*, ISTE Press, London and Elsevier, Oxford, 2016.

[DUR 16b] DUROUDIER J.-P., *Adsorption-Dryers for Divided Solids*, ISTE Press, London and Elsevier, Oxford, 2016.

[DUR 16c] DUROUDIER J.-P., *Solid–Solid, Fluid–Solid, Fluid–Fluid Mixers*, ISTE Press, London and Elsevier, Oxford, 2016.

[DUR 16d] DUROUDIER J.-P., *Solid–Solid, Fluid–Solid, Fluid–Fluid Mixers*, ISTE Press, London and Elsevier, Oxford, 2016.

[EL 84] EL-SHALL H., "Mechanisms of grinding modification by chemical additives: organic reagents", *Powder Technology*, vol. 38, pp. 267–273, 1984.

[ELI 94] ELIMELECH M., CHEN W.H., WAYPA J., "Measuring the zeta (electrokinetic) potential of reverse osmosis membranes by a streaming potential analyzer", *Desalination*, vol. 95, p. 269, 1994.

[ERG 52] ERGUN S., "Fluid flow through packed columns", *Chemical Engineering Progress*, vol. 48, no. 2, pp. 89–94, 1952.

[FAN 65] FANLO S., LEMLICH R., "Predicting the performance of foam fractionation columns", *AIChE Journal Symposium Series*, vol. 9, pp. 75–78, 1965.

[FLO 44] FLORY P.J., "Thermodynamics of heterogeneous polymers and their solutions", *Journal of Chemical Physics*, vol. 12, no. 11, p. 425, 1944.

[FLO 73a] FLOWER J.R., WHITEHEAD B.D., "Computer-aided design: a survey of flowsheeting programs – Part I", *Chemical Engineer (London)*, pp. 208–210, April 1973.

[FLO 73b] FLOWER J.R., WHITEHEAD B.D., "Computer-aided design: a survey of flowsheeting programs – Part II", *Chemical Engineer (London)*, pp. 271–277, May 1973.

[FUC 63] FUCHS N.A., STECHKINA I.B., "A note of the theory of fibrous aerosols filters", *Annals of Occupational Hygiene*, vol. 6, p. 27, 1963.

[FUE 70] FUERSTENAU R.W., HARPER R.W., MILLER J.D., "Hydroxamate vs. fatty acid flotation of iron oxide", *Transactions of the Society of Mining Engineers of A.I.M.E.*, vol. 247, pp. 69–73, 1970.

[GAL 71] GALE R.S., "Recent research on sludge dewatering", *Filtration and Separation*, vol. 8, pp. 531–538, 1971.

[GAR 96] GARDNER K.H., THEIS T.L., "A unified kinetic model for particle aggregation", *Journal of Colloid and Interface Science*, vol. 180, p. 162, 1996.

[GAV 77] GAVACH C., "Electrodialyse. Physico-chimie des membranes ioniques", *RGE*, vol. 86, no. 5, pp. 386–390, 1977.

[GOU 10] GOUY G., "Sur la constitution de la charge électrique à la surface d'un électrolyte", *Journal de Physique*, vol. 9, p. 457, 1910.

[GOU 17] GOUY G., "Sur la fonction électrocapillaire", *Annales de Physique*, vol. 7, p. 129, 1917.

[GOU 65] GOUDET G., *Les fonctions de Bessel*, Masson, 1965.

[GRA 56] GRACE H.P., "Structure and performance of filter media", *AIChE Journal*, vol. 2, p. 307, 1956.

[HAG 99] HAGMEYER G., GIMBEL R., "Modeling the rejection of nanofiltration membranes using zeta potential measurements", *Separation and Purification Technology*, vol. 15, p. 19, 1999.

[HAM 37] HAMAKER H.C., "The London–van der Waals attraction between spherical particles", *Physica IV*, no. 10, November 1937.

[HAY 72] HAYDON D.A., HADKY S.B., "Ion transport across thin lipid membranes: a critical discussion of mechanisms in selected systems", *Quarterly Reviews of Biophysics*, vol. 5, p. 187, 1972.

[HEL 79] HELMHOLTZ H., "Studien über electrische grenzschichten", *Annal der Physik und chemie*, vol. 7, pp. 337–382, 1879.

[HEN 07] HENDERSON P., "Zur Thermodynamik der Flussigkeitsketten", *Zeitschrift für Physik*, vol. 59, p. 118, 1907.

[HEN 08] HENDERSON P., "Zur Thermodynamik der Flussigkeitsketten", *Zeitschrift für Physik*, vol. 63, p. 325, 1908.

[HER 70] HERZIG J.-P., LECLERC D., LE GOFF P., "Review of filtration theories", *Industrial & Engineering Chemistry Research*, vol. 62, p. 8, 1970.

[HID 85] HIDALGO-ALVAREZ R., DE LAS NIEVES F.J., PARDO G., "Comparative sedimentation and streaming potential studies for ζ potential determination", *Journal of Colloid and Interface Science*, vol. 107, no. 2, p. 295, 1985.

[HOC 75] HOCHHAUSER M., CUSSLER E.L., "Concentrating chromium with liquid surfactant membranes", *Chemical Engineering Progress Symposium Series*, vol. 71, no. 152, p. 136, 1975.

[HOF 67] HOFFER E., KEDEM O., "Hyperfiltration in charged membranes: the fixed charge model", *Desalination*, vol. 2, pp. 25–39, 1967.

[HOG 66] HOGG R., HEALY T.W., FURSTENAU D.W., "Mutual coagulation of colloid dispersion", *Transactions of the Faraday Society*, vol. 62, p. 1638, 1966.

[HUG 41] HUGGINS M.L., "Solutions of long chain compounds", *Journal of Chemical Physics*, vol. 9, p. 440, 1941.

[IVE 78] IVES K.J., *The Scientific Basis of Flocculation*, NATO Advanced Study Institute Series, Sythoff & Nordhoff, 1978.

[JOH 85] JOHANSON J.R., "A rolling theory of granular solids", *Journal of Applied Mechanics*, vol. 32, no. 4, pp. 842–848, 1985.

[JOH 89] JOHANSON J.R., COX B.D., "Fluid entrainment effects in roll press compaction", *Powder Handling Processing*, vol. 1, no. 2, pp. 183–185, 1989.

[KAP 79] KAPLAN S.J., MORLAND C.D., HSU S.C., "Predict non-Newtonian fluid pressure drop across random-fiber filters", *Chemical Engineering*, pp. 93–98, 27 August 1979.

[KAR 67] KARJE B.H., "Permeability techniques for characterizing fine powders", *Powder Technology*, vol. 1, pp. 11–22, 1967.

[KAR 79] KARRA V.K., "Development of a model for predicting the screening performance of a vibrating screen", *CIM Bulletin*, vol. 72, pp. 167–171, 1979.

[KAT 91a] KATAOKA T., TSURU T., NAKAO S.-I. *et al.*, "Permeations equations developed for protection of membrane performance in pervaporation, vapour permeation and reverse osmosis based on the solution-diffusion model", *Journal of Chemical Engineering of Japan*, vol. 24, no. 3, pp. 326–333, 1991.

[KAT 91b] KATAOKA T., TSURU T., NAKAO S.-I. *et al.*, "Membrane transport properties of pervaporation and vapour permeation in ethanol–water system using polyacrylonitrile and cellulose acetate membranes", *Journal of Chemical Engineering of Japan*, vol. 24, no. 3, pp. 334–339, 1991.

[KEU 92] KEULEN (VAN) H., SMIT J.A.M., "Analytical approximations for potential profiles in charged micropores originating from the Poisson–Boltzmann equation", *Journal of Colloid and Interface Science*, vol. 151, no. 2, p. 546, 1992.

[KIM 84] KIMURA Y., LIM H.-J., IIJIMA T., "Membrane potentials of charged cellulosic membranes", *Journal of Membrane Science*, vol. 18, pp. 285–296, 1984.

[KIM 92] KIMURA S., "Transport phenomena in membrane separation processes", *Journal of Chemical Engineering of Japan*, vol. 25, no. 5, p. 469, 1992.

[KIN 74] KING R.P., HATON T.A., HUBBERT D.G., "Technical note. Bubble loading during flotation", *Translations of the Institution of Mining and Metallurgy*, vol. 83, no. 811, pp. C112–C115, 1974.

[KIN 75] KING R.P., "Simulation of flotation plants", *Translations of the Society of Mining Engineers of AIME*, vol. 258, pp. 286–293, 1975.

[KRO 98] KROL J.J., JANSINK M., WESSLING M. *et al.*, "Behaviour of bipolar membranes at high current density water diffusion limitation", *Separation and Purification Technology*, vol. 14, p. 41, 1998.

[KUM 67] KUMAR R., KULOOR N.R., "Blasen Bildung in Wirbelschichten", *Chemische Teknik – Berlin D.D.R.*, vol. 1, pp. 733, 1967.

[KUW 59] KUWABARA S., "The forces experienced by a lattice of elliptic cylinders in a uniform flow at small Reynolds numbers", *Journal of the Physical Society of Japan*, vol. 14, pp. 522–527, 1959.

[KYN 52] KYNCH G.J., "A theory of sedimentation", *Transactions of the Faraday Society*, vol. 48, p. 166, 1952.

[LAM 32] LAMB H., *Hydrodynamics*, 6th ed., Dover Publications, 1932.

[LEE 75] LEE C.H., "Theory of reverse osmosis and some other membrane permeation operations", *Journal of Applied Polymer Science*, vol. 19, pp. 83–95, 1975.

[LEE 78] LEE K.-H., EVANS D.F., CUSSLER E.L., "Selective copper recovery with two types of liquid membranes", *AIChE Journal*, vol. 24, p. 860, 1978.

[LEM 68] LEMLICH R., *Foam fractionation*, vol. 75, pp. 95–102, December 1968.

[LEO 65] LEONARD R.A., LEMLICH R., "A study of interstitial liquid flow in foam. Part I. Theoretical model and application to foam fractionation", *AIChE Journal*, vol. 11, no. 1, pp. 18–25, 1965.

[LEV 75] LEVINE S., MARRIOTT J.R., NEALE G. *et al.*, "Theory of electrokinetic flow in fine cylindrical capillaries at high zeta potentials", *Journal of Colloid and Interface Science*, vol. 52, no. 1, p. 136, 1975.

[LON 82] LONSDALE H.K., "The growth of membrane technology", *Journal of Membrane Science*, vol. 10, p. 81, 1982.

[LOV 66] LOVEDAY B.K., "Analysis of froth flotations kinetics", *Transactions of the Institution of Mining and Metallurgy. Section C. Mineral Processing and Extractive Metallurgy*, vol. 75, pp. C219–C225, 1966.

[LYK 68] LYKLEMA J., "Principles of the stability of lyophobic colloidal dispersions in non-aqueous media", *Advances in Colloid and Interface Science*, vol. 2, p. 65, 1968.

[MAC 55] MACKIE J.S., MEARES P., "The diffusion of electrolytes in cation-exchange resine membrane. I. Theroretical", *Proceedings of the Royal Society of London. Series A. Mathematical and Physical Sciences*, vol. A232, p. 498, 1955.

[MAR 78] MARZE X., "Progrès et développement de l'ultrafiltration", *L'Actualité Chimique*, pp. 37–42, September 1978.

[MAU 77] MAUREL A., "Electrodialyse. Procédés généraux de séparation par membranes (électrodialyse, osmose inverse, ultrafiltration). Procédés de séparation par membranes", *RGE*, vol. 86, no. 5, pp. 381–385, 1977.

[MCI 91] MCIVOR R.E., FINCH J.A., "A guide to interfacing of plant grinding and flotation operations", *Mineral Engineering*, vol. 4, no. 1, p. 9, 1991.

[MEY 36] MEYER K.H., SIEVERS J.F., "La perméabilité des membranes I – theorie de la perméabilité ionique", *Helvetica Chimica Acta*, vol. 19, pp. 649–664, 1936.

[MEY 37] MEYER K.H., SIEVERS J.F., "La perméabilité des membranes V sur l'origine des courants bioélectriques", *Helvetica Chimica Acta*, vol. 20, pp. 634–644, 1937.

[MIC 65] MICHAELS A.S., BIXLER H.J., HODGES R.M. JR., "Kinetics of water and salt transport in cellulose acetate reverse osmosis desalination membranes", *Journal of Colloid Science*, vol. 20, pp. 1034–1056, 1965.

[MOR 52] MORRIS T.M., "Measurement and evaluation of the rate of flotation as a function of particle size", *Transactions of the American Institute of Mining, Metallurgical and Petroleum Engineers Incorporated*, vol. 193, pp. 794–798, 1952.

[MUL 84] MULDER M.H.V., SMOLDERS C.A., "On the mechanism of separation of ethanol/water mixtures by pervaporation. Part I. Calculation of concentration profiles", *Journal of Membrane Science*, vol. 17, p. 289, 1984.

[MUL 85] MULDER M.H.V., FRANKEN A.C.M., SMOLDERS C.A., "On the mechanism of separation of ethanol–water mixtures by pervaporation. II Experimental concentration profiles", *Journal of Membrane Science*, vol. 23, pp. 41–58, 1985.

[NER 89a] NERNST W., "Die clektomotorische Wirksamkeit der ionen", *Zeitschrift für Physikalische Chemie*, vol. 4, p. 129, 1889.

[NER 89b] NERNST W., "Zur Kinetik der in Lösung befindlicher körper", *Zeitschrift für Physikalische Chemie*, vol. 2, p. 613, 1889.

[NOU 85] NOUGIER J.P., *Méthodes de calcul numérique*, Masson, 1985.

[NYS 94] NYSTRÖM M., PIHLAJAMÄKI A., EHSANI N., "Characterization of ultrafiltration membranes by simultaneous streaming potential and flux measurements", *Journal of Membrane Science*, vol. 87, p. 245, 1994.

[OBR 81] O'BRIEN R.W., HUNTER R.J., "The electrophoretic mobility of large colloidal particles", *Canadian Journal of Chemistry*, vol. 59, p. 1878, 1981.

[OPO 91] OPONG W.S., ZYDNEY A.L., "Diffusive and convective protein transport through asymmetric membranes", *AIChE Journal*, vol. 37, p. 1497, 1991.

[OST 35] OSTWALD W., "Electrolytkoagulation schwach solvatisierter Sole und Elektrolytactivität", *Kolloid Zeitschrift*, vol. 73, p. 301, 1935.

[OST 39] OSTWALD W., "Neuere Ergebnisse und Anschauungen über die Electrolytkoagulation hydrophober Sole", *Kolloid Zeitschrift*, vol. 88, p. 1, 1939.

[PAU 76] PAUL N.G., LUNDBLAD J., MITRA G., "Optimisation of solute separation by diafiltration", *Separation Science*, vol. 11, no. 5, pp. 499–502, 1976.

[PEE 99] PEETERS J.M.M., MULDER M.H.V., STRATHMANN H., "Streaming potential measurements as a characterization method for nanofiltration membranes", *Colloids and Surfaces. A. Physicochemical and Engineering Aspects*, vol. 150, p. 247, 1999.

[PIC 87] PICH J., "Gas filtration theory", in MATRESON M.J., ORR C. (eds), *Filtration Theory and Practice*, 2nd ed., Marcel Dekker, 1987.

[PLA 90a] PLANK M., "Ueber die Erregung von Eleklicität und Wärme in Elektolyten", *Annalen Der Physik Und Chemie*, vol. 39, p. 161, 1890.

[PLA 90b] PLANK M., "Ueber die Potentialdifferenz zwischen zwei verdünten Lösungen binärer Elektrolyte", *Annalen Der Physik Und Chemie*, vol. 40, p. 561, 1890.

[PLE 30] PLETTIG V., "Ueber die Diffusions potentiale", *Annalen der Physik*, vol. 5, p. 735, 1930.

[RAU 88] RAUTENBACH R., "Ultrafiltration of macromolecular solutions and cross-flow microfiltration of colloidal suspensions. A contribution to permeate flow calculations", *Journal of Membrane Science*, vol. 37, p. 231, 1988.

[REB 85] REBOUILLAT S., LECLERC D., BALUAIS G., "La déshydratation par filtration-pressage. Modélisation de la compression unidirectionnelle", *Entropie*, vol. 21, no. 121, pp. 13–29, 1985.

[RIC 54] RICHARDSON J.F., ZAKI W.N., "Sedimentation and fluidization: Part I", *Transactions of the Institution of Chemical Engineers*, vol. 32, p. 35, 1954.

[ROU 86] ROUX J.C., Thesis, National Polytechnic Institute of Grenoble, October 1986.

[RUM 85] RUMEAU M., "De la microfiltration et l'ultrafiltration à l'osmose inverse", *Informations Chimie*, no. 258, January–February 1985.

[SCH 50] SCHMID G., "Zur Electrochemic feinporiger kapillarsysteme. I. Uebersicht", *Zeitschrift für Elektrochemie*, vol. 54, no. 6, p. 427, 1950.

[SCH 53] SCHLÖGL R., "Ionenbeweglichkeiten in Austauschern", *Zeitschrift für Elektrochemie*, vol. 57, no. 3, pp. 195–201, 1953.

[SCH 54] SCHLÖGEL R., "Elektrodiffusion in freier Lösung und geladenen Membranen", *Zeitschrift für Physikalische Chemie (Neue Folge)*, vol. 1, pp. 305–339, 1954.

[SCH 74] SCHULTZ J.S., GODDARD J.D., SUCHDEO S.R., "Facilitated transport via carrier mediated diffusion in membranes", *AIChE Journal*, vol. 20, p. 417, 1974.

[SCH 77] SCHWARTZBERG H.G., ROSENAU J.R., RICHARDSON G., "The removal of water by expression", *AIChE Symposium Series*, vol. 73, no. 163, pp. 177–190, 1977.

[SCH 92] SCHAUM M.R., *Formules et tables de mathématiques*, McGraw Hill, 1992.

[SHE 65] SHERWOOD T.K., BRIAN P.L.T., FISHER R.E. *et al.*, "Salt concentration at phase boundaries in desalination by reverse osmosis", *Industrial and Engineering Chemistry Fundamentals*, vol. 4, no. 2, pp. 113–118, May 1965.

[SHI 79] SHIRATO M., MORASE T., ATSUMI K. *et al.*, "Industrial expression equation for semi-solid materials of solid–liquid mixture under constant pressure", *Journal of Chemical Engineering of Japan*, vol. 12, no. 1, pp. 51–55, 1979.

[SHI 86] SHIRATO M., MURASE T., IWATA M. *et al.*, "The Terzaghi–Voigt combined model for constant pressure consolidation of filter cakes and homogeneous semi-solid materials", *Chemical Engineering Science*, vol. 41, no. 12, pp. 3213–3218, 1986.

[SIG 75] SIGALÈS B., "How to design settling drums", *Chemical Engineering*, pp. 141–144, 23 June 1975.

[SIR 81] SIRKAR K., RAO G.H., "Approximate design equations and alternate design methodologies for tubular reverse osmosis desalination", *Industrial and Engineering Chemistry Design and Development*, vol. 20, pp. 116–127, 1981.

[SMO 03] SMOLUCHOWSKI M., "Contribution à la théorie de l'endosmose électrique et de quelques phénomènes corrélatifs", *Bulletin International de l'Académie des Sciences de Cracovie*, pp. 182–199, 1903.

[SMO 18] SMOLUCHOWSKI M., "Versuch einer mathematischen Théorie der Koagulationskinetik kolloïder Lösungen", *Zeitschrift für Physikalische Chemie*, vol. 92, p. 129, 1918.

[SOL 82] SOLES E., SMITH J.M., PARRISH W.R., "Gas transport through polyethylene membranes", *AIChE Journal*, vol. 28, no. 3, pp. 474–479, May 1982.

[SPI 66] SPIEGLER K.S., KEDEM O., "Thermodynamics of hyperfiltration (reverse osmosis). Criteria for efficient membranes", *Desalination*, vol. 1, p. 311, 1966.

[SPI 74] SPIEGEL R., *Formules et tables de mathématiques*, McGraw Hill, 1974.

[STE 24] STERN-HAMBURG O., "Zur Théorie der elektrolytischen Doppelschicht", *Zeitschrift für Elektrochemie*, vol. 30, p. 508, 1924.

[STE 66] STECHKINA I.B., "Diffuzionnoe osajdenie aerosolei v voloknistix filtrax", *Dokladi Akademii Nauk SSSR*, vol. 167, no. 6, pp. 1327–1330, 1966.

[STE 67] STECHKINA I.B., FUKS N.A., "Investigation of fibrous filters for aerosols. I. Calculation of diffusional deposition of aerosols in fibrous filters", *Colloid Journal of the USSR*, vol. 29, pp. 201–205, 1967.

[STE 69] STECHKINA I.B., KIRSCH A.A., FUCHS N.A., "Studies on fibrous aerosol filters IV. Calculation of aerosol deposition in model filters in the range of maximum penetration", *Annals of Occupational Hygiene*, vol. 12, p. 1, 1969.

[STR 04] STRATHMANN H., *Ion Exchange Membrane Separation Processes*, Elsevier, 2004.

[TAK 96] TAKAGI R., NAKAGAKI M., "Membrane charge of microporous glass membrane determined by the membrane potential method and its pore size dependency", *Journal of Membrane Science*, vol. 111, pp. 19–26, 1996.

[TAL 55] TALMAGE W.P., FITCH E.B., "Determining thickener unit areas", *Industrial & Engineering Chemistry*, vol. 47, no. 1, pp. 38–41, 1955.

[TEO 35] TEORELL T., "An attempt to formulate a quantitative theory of membrane permeability", *Proceedings of the Society for Experimental Biology and Medicine*, vol. 33, pp. 282–285, 1935.

[TEO 51] TEORELL T., "Zur quantitativer Behandlung der Membran-permeabilität", *Zeitschrift für Elektrochemie*, vol. 55, p. 460, 1951.

[TEO 53] TEORELL T., "Transport processes and electrical phenomena in ionic membranes", *Progress in Biophysics*, vol. 3, p. 305, 1953.

[TIL 77] TILLER F.M., CRUMP J.R., "Solid–liquid separation. An overview", *Chemical Engineering Progress*, vol. 73, p. 65, 1977.

[TIL 93] TILLER F.M., HSYUNG N.B., "How does percent solids affect centrifuge cakes", *Chemical Engineering Progress*, vol. 89, p. 20, 1993.

[TOY 67] TOYOSHIMA Y., KOBATAKE Y., FUJITA H., "Studies of membrane phenomena. Part 4. Membrane potential and permeability", *Transactions of the Faraday Society*, vol. 63, p. 2814, 1967.

[TRA 76] TRAHAR W.J., WARREN L.J., "The floatability of very fine particles – A review", *International Journal of Mineral Processing*, vol. 3, pp. 103–131, 1976.

[TRA 81] TRAHAR W.J., "A rational interpretation of the role of particle size in flotation", *International Journal of Mineral Processing*, vol. 8, pp. 289–327, 1981.

[TSU 90] TSURU T., NAKAO S.-I., KIMURA S., "Effective charge density and pore structure of charged ultrafiltration membranes", *Journal of Chemical Engineering of Japan*, vol. 23, p. 604, 1990.

[TSU 91] TSURU T., NAKAO S.-I., KIMURA S., "Calculation of ion rejection by extended Nernst–Plank equation with charged reverse osmosis membranes for single and mixed electrolyte solutions", *Journal of Chemical Engineering of Japan*, vol. 24, no. 4, p. 511, 1991.

[VER 48] VERVEY E.J.W., OVERBECK J.T.G., *Theory of the Stability of Lyophobic Colloids*, Elsevier, New York, 1948.

[VIS 72] VISSER J., "On Hamaker constants: a comparison between Hamaker constants and Lifschitz – Van der Waals constants", *Advances in Colloid and Interface Science*, vol. 3, p. 331, 1972.

[WAK 82] WAKEMAN R.J., "An improved analysis for the forced gas deliquoring of filter cakes and porous media", *Journal of Separation Processes Technology*, vol. 3, no. 1, pp. 32–38, 1982.

[WAN 95] WANG X.-L., TSURU T., NAKAO S.-I. *et al.*, "Electrolyte transport through nanofiltration membranes by the space-charge model and the comparison with Teorell–Meyer–Sievers model", *Journal of Membrane Science*, vol. 103, pp. 117–137, 1995.

[WAR 35a] WARK I.W., COX A.B., "Principles of flotation I. An experimental study of the effect of xanthates on contact angles at mineral surfaces", *Transactions of the American Institute of Mining and Metallurgical Engineers*, vol. 112, pp. 189–244, 1935.

[WAR 35b] WARK I.W., COX A.B., "Principles of flotation II. An experimental study of the influence of cyanide, alkalis and copper sulfate on the effect of potassium ethyl xanthate at mineral surfaces", *Transactions of the American Institute of Mining and Metallurgical Engineers*, vol. 112, pp. 245–266, 1935.

[WAR 35c] WARK I.W., COX A.B., "Principles of flotation III. An experimental study of the influence of cyanide, alkalis and copper sulfate on effect of sulphur-bearing collectors at mineral surfaces", *Transactions of the American Institute of Mining and Metallurgical Engineers*, vol. 112, pp. 267–302, 1935.

[WAR 70] WARD W.J., "Analytical and experimental studies of facilitated transport", *AIChE Journal*, vol. 16, p. 405, 1970.

[WER 95] WERNER C., JACOBASH H.-J., REICHELT G., "Surface characterisation of hemodialysis membrane based on streaming potential measurements", *Journal of Biomaterials Science, Polymer Edition*, vol. 7, no. 1, pp. 61–76, 1995.

[WOL 63] WOLVESPERGES A., "L'extraction de l'huile de palme au moyen des presses continues", *Oléagineux*, no. 11, pp. 717–723, 1963.

[WOO 71] WOODBURN E.T., KING R.P., COLBORN R.P., "The effect of particle size distribution on the performance of a phosphate flotation process", *Metallurgical Transactions*, vol. 2, no. 11, pp. 3163–3174, 1971.

[YAR 00a] YAROSHCHUK A.E., "Dielectric exclusion of ions from membranes", *Advances in Colloid and Interface Science*, vol. 85, p. 193, 2000.

[YAR 00b] YAROSHCHUK A.E., "Asymptotic behaviour in the pressure driven separation of ions of different mobilities in charged porous membrane", *Journal of Membrane Science*, vol. 167, p. 163, 2000.

[YU 77] YU L.T., MESSINA R., "Membranes échangeuses d'ions et membranes échangeuses d'électrons", *RGE*, vol. 86, no. 5, pp. 391–393, 1977.

Index

Printed in the United States
By Bookmasters

Printed in the United States
By Bookmasters